Validation of Chromatography Data Systems

Meeting Business and Regulatory Requirements

RSC Chromatography Monographs

Series Editor: R.M. Smith, *Loughborough University of Technology, UK*

Advisory Panel: J.C. Berridge, *Sandwich, UK*, G.B. Cox, *Indianapolis, USA*, I.S. Lurie, *Virginia, USA*, P.J. Schoenmakers, *Eindhoven, The Netherlands*, C.F. Simpson, *London, UK*, G.G. Wallace, *Wollongong, Australia*

Other titles in this series:

Applications of Solid Phase Microextraction
Edited by J Pawliszyn, *University of Waterloo, Waterloo, Ontario, Canada*

Capillary Electrochromatography
Edited by K D Bartle and P Myers, *University of Leeds, UK*

Chromatographic Integration Methods, Second Edition
N Dyson, *Dyson Instruments, UK*

Cyclodextrins in Chromatography
By T Cserháti and E Forgács, *Hungarian Academy of Sciences, Budapest, Hungary*

HPLC: A Practical Guide
T Hanai, *Health Research Foundation, Kyoto, Japan*

Hyphenated Techniques in Speciation Analysis
Edited by J Szpunar and R Lobinski, *CNRS, Pau, France*

Packed Column SFC
T A Berger, *Hewlett Packard, Wilmington, Delaware, USA*

Separation of Fullerenes by Liquid Chromatography
Edited by *Kiyokatsu Jinno, Toyohashi University of Technology, Japan*

Electrochemical Detection in HPLC: Analysis of Drugs and Poisons
R J Flanagan, *Guy's and St Thomas' NHS Foundation Trust, London, UK*, D Perrett, *Queen Mary's School of Medicine and Dentistry, London, UK* and R Whelpton, *University of London, London, UK*

How to obtain future titles on publication

A standing order plan is available for this series. A standing order will bring delivery of each new volume upon publication. For further information please contact:

Sales and Customer Care
Royal Society of Chemistry, Thomas Graham House
Science Park, Milton Road, Cambridge, CB4 0WF
Telephone: +44(0) 1223 420066, Fax: +44(0) 1223426017, Email: sales@rsc.org

RSC
CHROMATOGRAPHY
MONOGRAPHS

Validation of Chromatography Data Systems

Meeting Business and Regulatory Requirements

R.D. McDowall
McDowall Consulting, Bromley, Kent, UK

advancing the chemical sciences

ISBN 0-85404-969-X

A catalogue record for this book is available from the British Library

Published by The Royal Society of Chemistry,
Thomas Graham House, Science Park, Milton Road,
Cambridge CB4 0WF, UK

Registered Charity Number 207890

For further information see our web site at www.rsc.org

Typeset by Alden Bookset, Northampton, UK
Printed by Athenaeum Press Ltd, Gateshead, Tyne and Wear, UK

Preface

Why read or even buy this book?

If you are using a chromatography data system (CDS) in the regulated areas of the pharmaceutical, medical device, active pharmaceutical ingredient and contract research organisations, you will need to validate the system.

This book will be your guide through the regulations and jargon. It provides practical advice that can be used directly by you to meet regulatory requirements and allow a sustainable validation effort for your chromatography data system throughout its operational life.

However, computer validation is more than just a means of meeting regulatory requirements. It is a strategic business tool.

- How much money has your organisation wasted on computer systems that fail to meet initial expectations or do not work? If used correctly, validation is a means of implementing the right system for the right job. Computer validation is quite simply good business practice, that, if followed provides regulatory compliance for no additional cost.
- In addition, implementing electronic signatures with electronic ways of working will allow a laboratory to exploit tangible business benefits from regulatory compliance. This requires more time spent mapping and analysing the current working process and practices but the payback is reduction of tedious tasks such as checking for transcription errors in the laboratory and tangible time and resource savings.

This book is intended to help the reader to validate their CDS in the current risk based regulatory climate and is written by a chromatographer with extensive experience of validating many different computerised systems in many different organisations since 1986.

The principles and practices of validation outlined in this book are also applicable to other types of computerised systems used in laboratories.

Bob McDowall C.Chem. FRSC
Principal, McDowall Consulting
Bromley, UK

Acknowledgements

I would like to thank the following for their review comments during the preparation of this book: Chris Burgess, Sava Lukac and Mark Mercer. Waters Corporation kindly supplied the screen shots for Figures 25 and 26.

All trademarks used within the text are acknowledged.

Contents

List of Figures

List of Tables

Glossary

Term	Meaning
Acceptance criteria	The criteria a system must meet to satisfy a test or other requirement (PDA Technical Report 18)
Boundary value	A minimum or maximum input, output or internal data value, applicable to a system (PDA Technical Report No. 18)
Branch testing	Execution of tests to assure that every branch alternative has been exercised at least once
Change control	A formal monitoring system by which qualified representatives of appropriate disciplines review proposed or actual changes that might affect a validated status to determine the need for corrective action that would assure that the system retains its validated state
COTS Software	Commercial off the shelf software applications that are used as is or can be configured to specific user applications by "filling in the blanks" without altering the basic program
Commissioning	Activities to put a computerised system into proper operation
Computer hardware	Various hardware components in the computer system, including the central processing unit, printer, screen and other related apparatus
Computer system	Computer hardware components assembled to perform in conjunction with a set of programs, which are collectively designed to perform a specific function or group of functions
Computer related system	One or more computerised systems and the relevant operating environment
Computerised system validation	Establishing documented evidence which provides a high degree of assurance that a specific computer related system will consistently operate in accordance with predetermined specifications

Computerised system	A computer system plus the controlled function which it operates
Computerised system specification	A document or set of documents that describe how a computerised system will satisfy the system requirements to the computer related system
Controlled function	A process and any related equipment controlled by a computer system
Design specification	A specification that defines the design of a system or system subcomponents
Developer	The company or group responsible for developing a system or some portion of a system
Firmware	A software program permanently recorded in a hardware device such as a chip or EPROM
Functional requirements	Statements that describe functions, a computer related system must be capable of performing
Functional specification	Statements of how the computerised system will satisfy functional requirements of the computer related system
Functional testing	A process for verifying that software, a system or a system component performs its intended functions
Installation Qualification (IQ)	Documented verification that all key aspects of hardware and software installation adhere to appropriate codes and the computerised system specification
Ongoing evaluation	The dynamic process employed after a system's initial validation to maintain its validated state
Operating environment	Those conditions and activities interfacing directly or indirectly with the system of concern, control of which can affect the system's validated state
Operating system	A set of software programs provided with a computer that function as the interface between the hardware and the applications program
Operational Qualification (OQ)	Documented verification that the system or subsystem operates as specified in the computerised system specifications throughout representative or anticipated operating ranges
Path testing	Execution of important control flow paths throughout the program
Performance Qualification (PQ)	Documented verification that the integrated computerised system performs as intended in its normal operating environment, *i.e.* that computer related system performs as intended
Policy	A directive usually specifying what is to be accomplished
Process	Structured activities intended to achieve a desired outcome

Qualification protocol	A prospective experimental plan that when executed is intended to produce documented evidence that a system or subsystem has been qualified properly
Standard Operating Procedure	Instructions that specify how something is to be accomplished
Structural integrity	Software attributes reflecting the degree to which source code satisfies specified software requirements and conforms to contemporary software development
Structural verification	An activity intended to produce documented assurance that software has the appropriate structural integrity
User	The company or group responsible for the operation of the system
Validation plan	A document that identifies all systems and subsystems involved in a specific validation effort and the approach by which they will be qualified and the total system will be validated; includes the identification of responsibilities and expectations
Vendor	The company or group responsible for developing, constructing and delivering a system or part of a system

Abbreviations

Abbreviation	Meaning
ANSI	American National Standards Institute
API	Active Pharmaceutical Ingredient
ASTM	American Society for Testing and Materials
CANDA	Computer Assisted New Drug Application
CBER	Center for Biologics Evaluation and Research
CDER	Center for Drug Evaluation and Research
CDRH	Center for Devices and Radiological Health
CDS	Chromatography Data System
CFR	Code of Federal Regulations (*e.g.* 21 CFR 11)
CPG	Compliance Policy Guide
CRO	Contract Research Organization
cGMP	Current Good Manufacturing Practice
COA	Certificate of Analysis
COTS	Commercial Off The Shelf
CSV	Computer System Validation
DHHS	Department of Health and Human Services (US)
DQ	Design Qualification
DS	Design Specification
EC	European Community
EDMS	Electronic Document Management System
EIR	Establishment Inspection Report
EMEA	European Agency for the Evaluation of Medical Products
EP	European Pharmacopoeia (Ph.Eur.)
EU	European Union
FDA	Food and Drug Administration
FMEA	Failure Mode and Effects Analysis
FOI (A)	Freedom of Information (Act)
FR	Federal Register
F(D)S	Functional (Design) Specifications
GAMP	Good Automated Manufacturing Practice guidelines
GC	Gas Chromatography
GERM	Good Electronic Record Management

GLP	Good Laboratory Practice
GXP	Good X Practices (clinical, laboratory or manufacturing)
GMP	Good Manufacturing Practice
HACCP	Hazard Analysis Critical Control Point
HPLC	High Pressure Performance Liquid Chromatography
ICH	International Conference on Harmonisation
IND	Investigational New Drug Application
IEEE	Institute of Electrical and Electronic Engineers, Inc.
IQ	Installation Qualification
ISO	International Organization for Standardization
ISPE	International Society of Pharmaceutical Engineers
LAN	Local Area Network
LIMS	Laboratory Information Management System
MHRA	Medicines and Healthcare products Regulatory Agency (UK)
MHLW	Ministry of Health, Labour and Welfare (Japan)
MOU	Memorandum of Understanding
MRA	Mutual Recognition Agreement
NCE	New Chemical Entity
NDA	New Drug Application
NIST	National Institute of Standards and Technology (Gaithersville, Maryland, USA)
OECD	Organization for Economic Cooperation and Development
OOS	Out of Specification
OOT	Out of Trend
OQ	Operational Qualification
ORA	Office of Regulatory Affairs (FDA)
PAI	Pre-Approval-Inspection
PDA	Parenteral Drug Association
PDF	Portable Document Format
Ph.Eur.	European Pharmacopoeia
PIC	Pharmaceutical Inspection Convention
PIC/S	Pharmaceutical Inspection Convention/Scheme
PKI	Public Key Infrastructure
PQ	Performance Qualification
QA	Quality Assurance
QAU	Quality Assurance Unit
QC	Quality Control
QMS	Quality Management System
R&D	Research and Development
SDLC	System Development Life Cycle
SLA	Service Level Agreement
SOP	Standard Operating Procedure
SQA	Society for Quality Assurance

SRS	System Requirements Specification (equivalent to URS)
UPS	Uninterruptible Power Supply
URS	User Requirement Specifications
USP	United States Pharmacopoeia
VMP	Validation Master Plan
WAN	Wide Area Network

CHAPTER 1

How to Use this Book

1.1 Purpose and Scope

Chromatography is a major analytical technique that is used in almost all analytical laboratories. The days of chart recorders and paper and pencil interpretation have gone and today the chromatography data generated by a method is now acquired, stored, interpreted, manipulated and reported by a chromatography data system (CDS).

When a laboratory operates in a controlled industry, such as the pharmaceutical, biotechnology or medical device along with the allied contract research organisations, the applicable regulations require that the CDS be validated for its intended purpose. However, in today's world where many organisations work in a global market, there are many regulations that are applicable even within a single laboratory.

The purpose of this book is to give readers a practical understanding of how to validate their CDS. The principles outlined here are applicable from single standalone systems to client server systems for a site and to larger terminal served systems operating between sites and over two or more time zones. The reader needs to scale the principles in this book to their specific system and ways of working.

Analytical laboratories working in analytical research and development and manufacturing under Good Manufacturing Practice (GMP) regulations as well as bioanalytical laboratories operating under Good Laboratory Practice (GLP) regulations can use this book. This book also includes validation of mass spectrometry data systems used for quantitative analysis in a regulated environment.

In this book, I want to discuss the prospective validation of CDS software. By prospective validation, I mean undertaking the validation work in parallel with progress through the life cycle of the project from start to finish. Unfortunately, this is not always the case. Usually just before the system goes live someone thinks that perhaps we should validate the system! Taking this approach will add up between 25 and 50% to the validation costs of the project. The main reason is documentation that should have been written at key stages of the project is missing or if written may not be of sufficient quality for laboratories working under regulations such as GMP or GLP. However, some people may approach CDS validation retrospectively and in Chapter 25 there is an outline of what should be done in this situation. However, the main emphasis in this book is on prospective validation.

1.2 The Way It Was. . .

In the past, the chromatograph and CDS software was purchased and then just before it was put into operational use someone thought about validation of the system. Some common questions may have been:

- Have we validated the system? No
- Does it matter? Probably
- Will we get caught? Do not even think about answering no to this question

Considering validation at such a late stage of the life cycle will mean a delay in going live, thus failing to gain benefit from the investment in the instrument and releasing the system with no regulatory coverage. This depends on your approach to risk and if can you sleep at night.

This approach to validation had no concept or consideration of a system development life cycle (SDLC) or even testing the system to see if it was capable of supporting the laboratory.

1.3 The Way It Should Be. . .

However, a proactive approach to validation is necessary and if done correctly will actually save you money by ensuring that you buy the right CDS for your laboratory to meet the defined and intended role of the system. So we will start at the beginning and look at the first stages of the SDLC (a defined life cycle is one of the foundations of computer validation that will be discussed in more detail in Chapter 4):

- Defining and controlling the validation throughout the whole life cycle (writing the validation plan)
- Specifying what you want the system to do (writing a user requirements specification, URS)
- Selecting the system using the requirements defined in the URS on an objective basis rather than a glossy brochure.

1.4 Book Structure: Life to Death of a CDS

The structure of this book is presented graphically in Figure 1. It contains seven phased themes with the remaining chapters divided amongst them that cover the complete life cycle of a CDS. Each will be described in more detail in the remaining sections of this introductory chapter. You will find this figure a useful starting point when starting or returning to this book. Figure 2 shows how the chapters link with the process for specifying a CDS through to when the system first goes live within a laboratory. Figure 3 shows the chapters related to maintaining the validation of the system throughout its operational life and into system retirement.

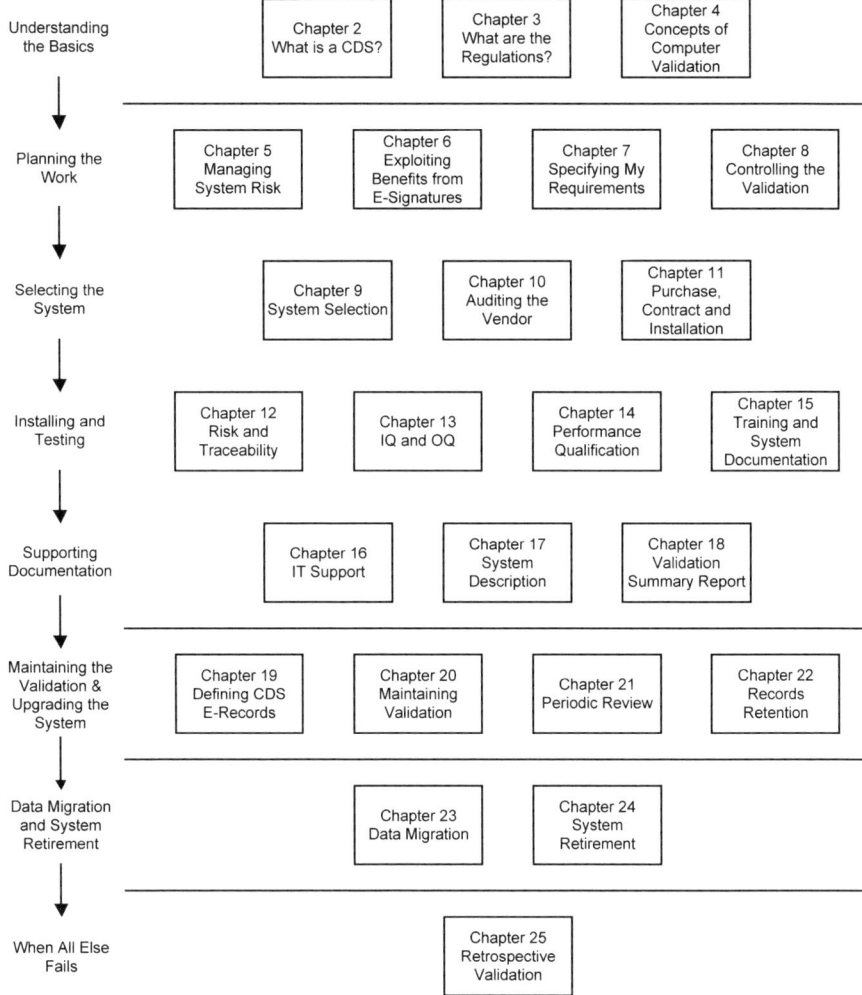

Figure 1 *Outline chapter structure of this book*

1.4.1 Chapter Structure

The majority of chapters in this book are written in the same way:

- The chapter starts with a brief overview why the chapter is important within the overall scheme of CDS validation.
- This is followed by a section of regulatory requirements that are relevant to the chapter; thereby positioning the regulatory rationale for what you are doing.
- Where appropriate, there is also the business rationale for the tasks contained in the chapter.
- Then there is a discussion of how to achieve the objective of each chapter. For example, if you are writing the URS, how this can be achieved and how to

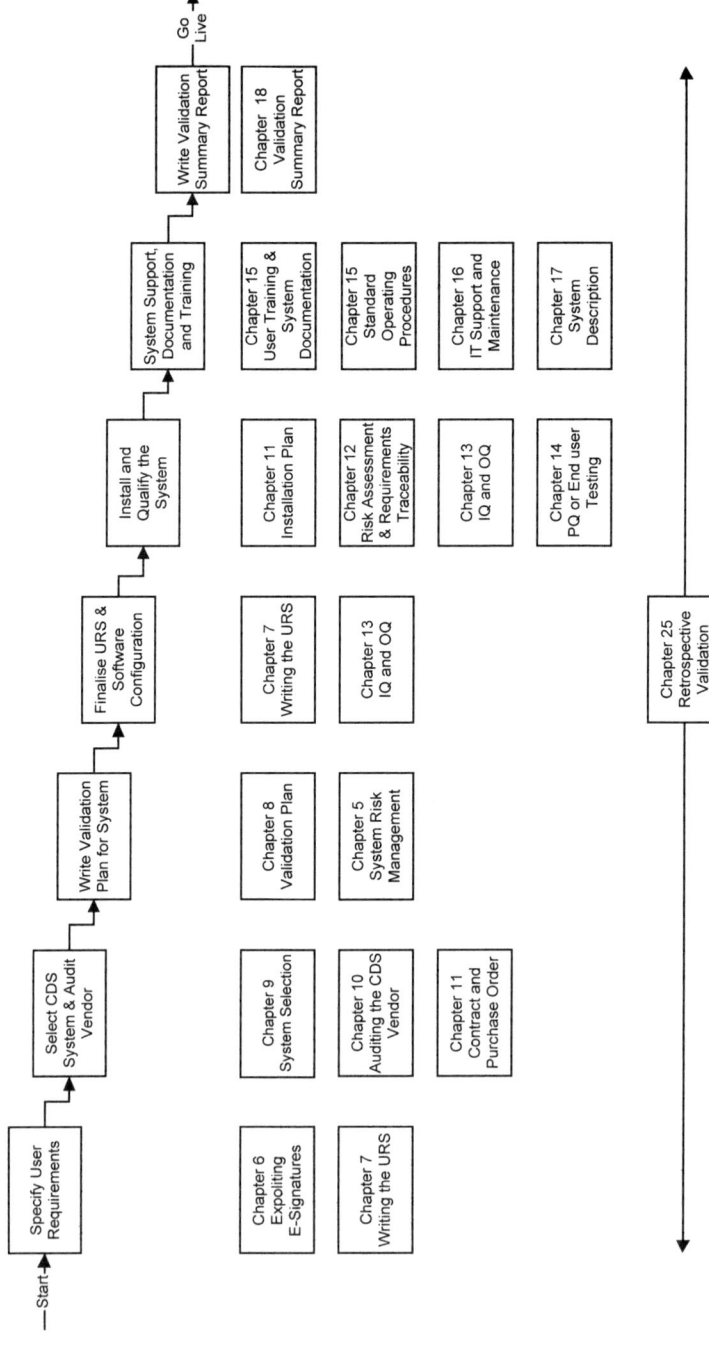

Figure 2 *Process for a CDS validation from start to go live and linked book chapters*

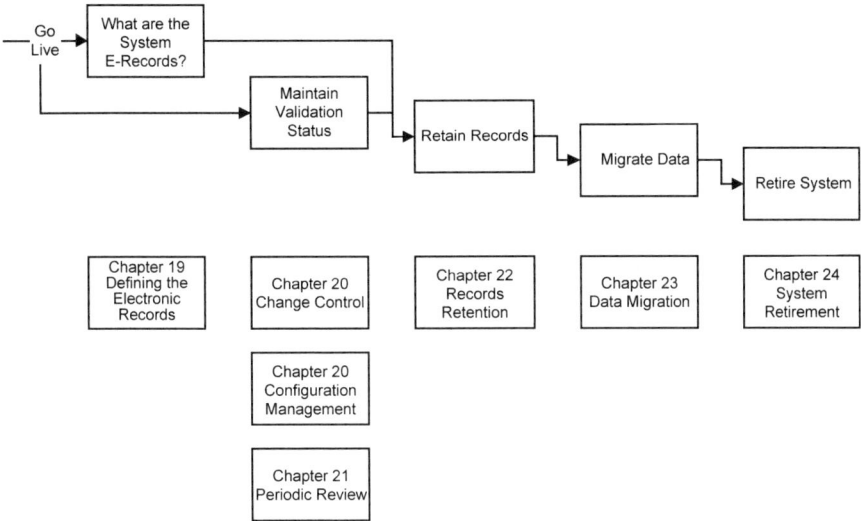

Figure 3 *Maintaining the validation and system retirement mapped to the book chapters*

avoid some of the common pitfalls. The aim is to give any reader the practical basis and confidence to work on any subject covered by this book.

The intention of this approach is to put the regulatory and business rationale for performing a task at the reader's fingertips. It also allows an individual chapter to stand alone if a quick reference on a specific topic is all that is required.

1.4.2 Understanding the Basics

Chapters 2–4 are used to introduce the topic of CDS validation and set the scene for the remainder of the book:

- Introduction to CDS
- Regulatory requirements for CDS validation
- Key terms and concepts of computer validation

If you are new to the subject, these chapters are intended to give you an understanding of the topics and to lead to further reading if necessary.

1.4.3 Planning the Work

Planning any validation is critical and Chapters 5–8 cover, respectively:

- CDS validation: managing risk
- Exploiting the business tangible benefits of electronic signatures with a CDS
- Writing the URS
- Controlling the validation: the validation plan

The first question to be asked is do I need to validate the system or not, therefore we start our validation journey by asking this fundamental question. Once decided we need to plan the work, this is the foundation for the overall quality of the validation. Quality is designed and not tested into a system. The whole process must be controlled and a clear idea of what will be expected at the end of the validation is documented before the real work starts.

Understanding the process is an important part of implementing and validating a new CDS – with or without the use of electronic signatures or implementing electronic signatures in a new version of an application. Therefore, we consider this question early in the book. The choice of writing the URS before the validation plan reflects the practical situation found in many laboratories, the requirements are written before selecting the system which will then be validated. Therefore, the URS comes before the validation plan in this book.

1.4.4 Selecting the System

Chapters 9–11 cover the system selection phase:

- System selection
- Auditing the CDS vendor
- Contract, purchase order and planning the installation

The aim of these chapters is to have the right tool for the right job that has been correctly developed by the right vendor. Make sure that there is sufficient emphasis at this stage of the life cycle as once the system has been purchased there will be no opportunity to change for a long time.

If your CDS has already been purchased and you are validating an upgrade to the system, then this section of the book can be omitted if required.

1.4.5 Installing and Testing the System

A major part of the life cycle before the system goes live is the installation and testing of the system covered in Chapters 12–15:

- Risk assessment and requirements traceability
- Installation qualification and operational qualification (IQ and OQ)
- Performance qualification (PQ) or end-user testing
- Training and system documentation

Deciding what user requirements to test and where they are tested is based on risk assessment and traceability matrix, respectively. Then the system components must be correctly installed and work as described in the URS to demonstrate fitness for intended purpose. The use of a CDS vendor's material for IQ and OQ must be assessed critically to see the value for money that you will be obtaining and if you can reduce any PQ testing if the vendor material matches your written requirements.

There is a detailed discussion of what is PQ or end-user testing and how to design and execute test procedures. A vital part of the validation is to ensure all users are correctly trained and there are written procedures available to follow.

1.4.6 Support and Release of the System

Three chapters (16–18) discuss the supporting documentation required before the system goes live as well as what you do if you have an existing system to validate (Chapter 19):

- IT support and maintenance
- System description
- Validation summary report

Critical functions of all CDS systems will require backup and restore plus other IT functions; how are these controlled? As a system description is required for regulatory reasons, a chapter provides how this is can be written. The whole validation effort is reported in a summary report along with a release statement signed by the system owner.

1.4.7 Maintaining the Validation and Upgrading the System

The easy job is over and the biggest validation challenge remains: maintaining the validation throughout the operational life of the system which may be many years. Chapters 19–22 present:

- Defining electronic records for a CDS
- Maintaining validation status during operational life
- Records retention

In preparation for inspections, how can you define and document the electronic records that your CDS will produce and also the steps you must take to maintain the validation status through its operation life. The various options for retaining the records produced by a CDS are discussed.

1.4.8 Data Migration and System Retirement

The final stages of the life cycle of a CDS are data migration and system retirement; these topics are covered in Chapters 23 and 24:

- CDS data migration
- System retirement

How critical are your CDS data? If your system is obsolete and needs to be retired you need to migrate the data to a new system or select an alternative approach. Afterwards the components of the system are formally retired.

1.4.9 Retrospective Validation of a CDS

Existing CDS applications operating in regulated laboratories that have not been validated will need to comply with regulations retrospectively. As the concept of computer validation in the pharmaceutical industry has been discussed since the early 1980s, there should not be many systems that fall into this category. However, from experience this is not the case. Therefore, the final chapter in this book briefly covers retrospective validation and shows how to perform a gap and plan analysis and links back to the other chapters in this book for the detail of how to carry out the remedial activities.

1.4.10 Use Your Organisation's Computer Validation Policy

The approaches outlined in this book need to be tailored to fit with the computer validation policy of the reader's organisation. Some organisations are more conservative than others and therefore more work will be done than outlined here. In contrast, some companies may want to do less than I present in the following chapters. The choice is yours. Computer validation has some elements that are given and are not open for discussion. In other areas, there is a degree of interpretation; is my interpretation closer or further away from yours?

1.5 Assumptions, Exclusions and Limitations

Owing to the size and scope of this book, there are some of the assumptions, exclusions and limitations of what can be covered in the validation of CDS.

- I assume that your organisation has a corporate computer validation policy available. You will need to interpret the approach outlined here for your specific corporate requirements. We discuss the principles of a CSV policy in Sections 1.4.10 and 4.8 but do not discuss the detail of computer validation policies or validation master plans any further.
- Network infrastructure is assumed to be qualified and under control within your organisation and therefore will not be covered. The exception is Chapter 13 where the IQ and OQ of the CDS database server is mentioned briefly.
- In this book, I make few references to the Good Automated Manufacturing Practice (GAMP) guidelines apart from Appendices M3 and M4 (risk assessment and validation strategies for different classes of software, respectively).[1] The issue is that GAMP provides an overall framework for validation of automated manufacturing equipment as well as software applications. Note here the use of the word "framework". In my opinion, there is not enough detail in GAMP to really provide sufficient practical advice to help you validate a CDS. Institute of Electronic and Electrical Engineers (IEEE) software engineering standards,[2] on the other hand provide much more detail and therefore selected and relevant IEEE software engineering standards that are referenced in preference, where appropriate, to GAMP within this book.
- The chromatographic equipment interfaced to a CDS will not be discussed in detail and is assumed to be qualified and working correctly.

Introduction to Chromatography Data Systems

Chromatography is an analytical technique used in virtually all sectors of the pharmaceutical, medical device and biotechnology industries to detect or quantify compounds during the course of product development and manufacture. It can be used for the assessment of active ingredients, raw materials, impurities and determining the stability of active in final preparations. The chromatograms generated by these analytical methods are displayed, integrated and results calculated by a software application called a chromatography data system (CDS).

2.1 What is a Chromatography Data System?

This section discusses the operation of a CDS from the perspective of a typical laboratory process or workflow. Figure 4 shows the main features of a CDS and Figure 5 shows the overall sequence of events that a typical data system should perform. This is a generalised approach to the operation of a "typical" data system; further details on the subject are the Royal Society of Chemistry monograph by Dyson,[3] the book by Fellinger on Data Analysis and Signal Processing in Chromatography[4] and in the chapter and articles by McDowall.[5,6]

This understanding is important as detailed knowledge of how a specific CDS application works is essential to write and maintain a user requirements specification throughout the operational lifetime of any system.

2.1.1 Types of Chromatography Data System

A CDS can come in one of the following types:

- integrator (single user and single instrument data acquisition)
- workstation (typically a single user with single instrument data acquisition and control)
- client–server (multiple user and multiple instrument data acquisition with an option for control)

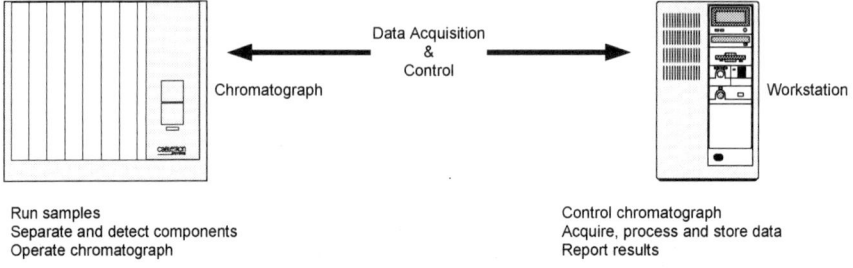

Figure 4 *Outline chromatograph and CDS*

- terminal server (multiple user *via* a terminal server farm with multiple instrument data acquisition with an option for control)

The essential elements of a CDS and how it can interact with the gas or liquid chromatograph are shown diagrammatically in Figure 4:

- instrument control
- data acquisition typically *via* an analogue to digital (A/D) converter or direct through a digital interface
- integration of the data
- calculation and reporting of results

This chapter is a general introduction to the main functions of a CDS. Chapter 19 discusses the definition of the electronic records (raw data) and metadata associated with a system in more detail.

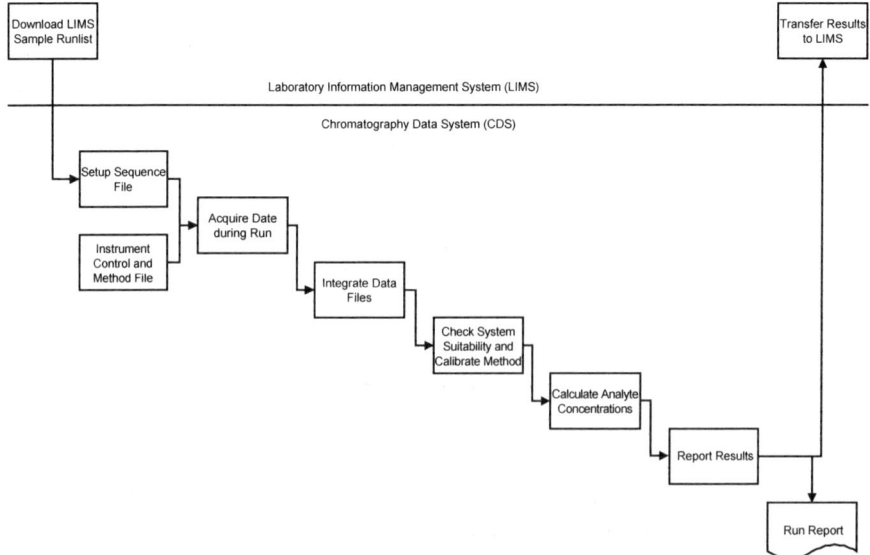

Figure 5 *Process workflow for a typical chromatography data system with optional inter-face to a LIMS*

2.1.2 Naming Conventions

Before starting a discussion of the functions of a CDS, it is important to understand the need for naming conventions used within a system for method, instrument control, sequence, reports and all data files stored within it. This is to avoid having files named the same or in an unusual manner that makes retrieval difficult. Any CDS must have sufficient capacity for naming all the files that would be created by the system over the records retention for the regulations it operates under. This will aid efficient storage, archiving and unambiguous identification of these files for easier retrieval. Therefore, for efficient management of data files and methods, naming conventions must be introduced. Any naming convention system must aid users, quality assurance and regulatory inspectors.

A naming convention should be based on the workflow undertaken by a laboratory. This is to allow not only efficient archiving of data but also, just as importantly, the efficient retrieval of data. Some ideas might be:

- Organise the data around drug products or development projects, as this is often how the work is structured and how many project teams are organised. This will help retrieve data to aid 21 CFR 11 compliance for ready retrieval of electronic records.
- Create major subdivisions of each project based around the type of work done, *e.g.* method development, method validation, pre-formulation, *etc.*

2.1.3 Method Files

The start of the data acquisition operation of a CDS is to build a method file. This tells the data system how to acquire data and process and interpret the results. A method file should control:

- the data sampling rate of the analogue to digital (A/D) converter[7]
- when to start and stop the integration of the chromatogram
- whether peak areas or heights should be used
- retention time windows and identification of the analytes and internal standard
- allocate the method to calculate the analyte amount or concentration

A name, number or a mixture of both should identify individual method files within the system. In addition, the system should be able to provide facilities for version control of method files to ensure that control is maintained over the method for the lifetime of its use. Part of the control function must be access control to identify the individuals who can create, modify or delete analytical methods. If a method has been modified, then copies of the modifications must be stored with the data processed by that method. This is to provide an audit trail for the data and results produced by a version of a method. However, when developing methods, flexibility with method files is essential and a default method should be available to acquire data and then feedback to a normal method.

2.1.4 Instrument Control Files

The primary interaction of the CDS with analytical instrumentation is with the output from the detector. However, there are other considerations such as instrument set-up and control. These can vary from system to system and the following options are available:

- Contact closures for the control of chromatographic valves or associated equipment during analysis is usually available for other vendor's equipment.
- When the same vendor makes the data system and the chromatography equipment, control is more sophisticated and more tightly integrated with the data system functions so control of the instrument and set-up of the data system can be achieved from a single workstation. Other CDS vendors can develop control codes for other vendor's chromatographic equipment or have joint agreements to use them to achieve the same aims.
- Communication with the auto-sampler *via* binary-coded decimal (BCD) or equivalent communication for sample continuity is, in my view, essential but is usually ignored by many laboratories and only offered as an option by most vendors.
- Remote monitoring of the chromatography system output including the instrument conditions.

Some CDS systems can also list the items of equipment (pump, detector, *etc.*) used for a particular analysis, which is a useful function that helps to automate the administrative records associated with an analysis and help meet GXP compliance. The ability of a CDS to control and monitor any changes to the instruments being controlled is an advantage within a regulated environment. When a parameter is changed during an analytical run such as an increase or decrease of the flow rate of the mobile phase or carrier gas, this should be accompanied by an audit trail entry explaining the reason for the change.

2.1.5 Sequence File

The sequence file is the run list or order that the samples, standards, quality control samples and blanks will be injected into the chromatograph. This is essential as it puts into context the content of the individual data files. Each injection within a sequence file must be linked to a specific method file to process the resulting data. For laboratories with large numbers of samples for a single method, the sequence file will usually be linked with a single method. Smaller laboratories may need the flexibility to link the sequence file with several methods during the course of a single analytical run for best use of equipment resources or to flush the chromatographic column and possibly turn off the instrument.

Each sample to be analysed should be identified in the sequence file as one of the following types:

- unknown
- calibration standard
- quality control
- blank

Depending on the data system involved, at least the first two options are available to a user. There may also be a sample number to link the injection to the physical sample used for analysis.

Sample identities can either be typed into the sequence file directly by the user or the information can be downloaded electronically from a Laboratory Information Management System (LIMS) as shown in Figure 5.

2.1.6 Acquisition of Chromatographic Data

The signal from the detector needs to be acquired by the CDS. This can occur in one of the two ways. The traditional way is *via* an analogue to digital (A/D) data converter. A/D conversion is a process by which a continuously variable signal (*e.g.* detector voltage) is converted to a binary number that can accurately represent the original data. It is necessary to convert the analogue signal to a digitised form because computer systems only handle numerical information in the form of a binary number. A detailed discussion of the principles of A/D conversion is outside the scope of this book and the reader is referred to the book by Dyson[3] or the article by Burgess *et al.*[7] For a discussion of more technical details of the CDS such as data collection rates, bunching factors and slope sensitivity, the reader is referred to Dyson's book.[3]

Alternatively, some chromatographs have digital data acquisition where the instrument output can be input directly to a data server and can be manipulated without the need to use an external or internal A/D unit.

To ensure the trustworthiness and reliability of the data generated by the system, vendors incorporate a checksum for each data file; if any changes have been made to the file, the checksum will be wrong and the file will be unable to be opened by the application.

2.1.7 Management of Data: Database or Files?

Chromatographic data and associated files such as method, sequence, results can be stored in one of the two main ways: files within the operating system directories or within a database. There are many advantages within a regulated environment for the use of a database. The main reason for this statement is the disadvantage of the directories within the operating system. Users can access files outside of the application by using operating system utilities such as Windows Explorer to delete files without any audit trail entry. In contrast, a database can automatically manage version control of method files and sequences without operator input if the software application has been designed appropriately.

2.1.8 Interpretation of Chromatographic Data

After the method file and the sequence file have been set up, the analytical run is started and data are collected. A data file containing the A/D data slices will be obtained for each chromatographic run and sample injected, it will be plotted as illustrated in Figure 6. It is important from scientific and regulatory considerations

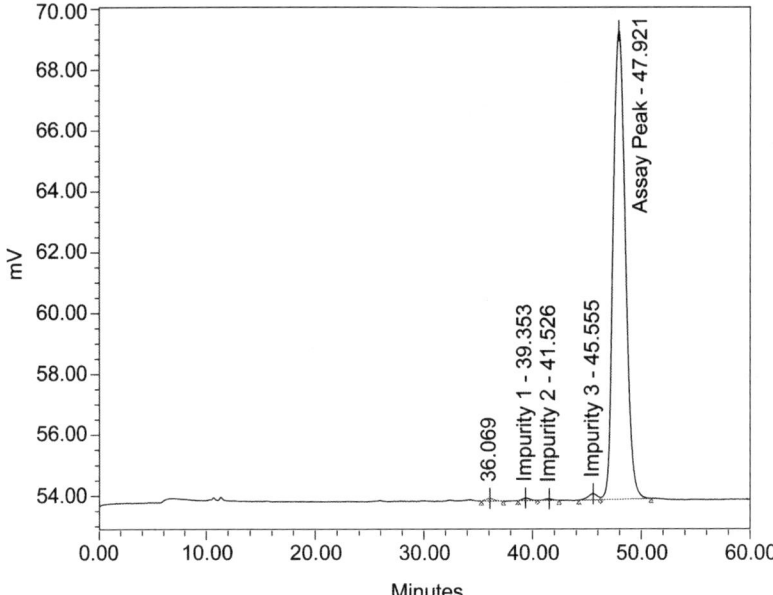

Figure 6 *A typical chromatogram of an active substance separation from impurities and degradation products*

that the data files must not be capable of alteration and a vendor will usually incorporate a mathematical checksum in the data file to identify any tampering.

Moreover, the data files must not be overwritten. This is a key area for consideration when validating the CDS; as you must know what happens to your data files, especially in a regulated environment.

The data system will interpret each data file, identifying the individual peaks and fitting the peak baselines according to the parameters defined in the CDS method. The data systems should have the ability to identify whether the peak baselines have been automatically or manually interpreted. This is a useful feature for compliance with Part 11 to indicate the number of times a chromatogram has been interpreted.

Most data systems should be able to provide a real-time plot, so that the analyst can review the chromatograms as the analytical run progresses. In addition, the plotting options of a data system should include:

- fitted baselines
- peak start/stop ticks
- named components
- retention times
- timed events, *e.g.* integration start/stop
- run-time windows and user-defined plotting windows
- baseline subtract

Each of these options should be capable of being enabled or disabled by a user.

An overlay function should be available to enable you to compare results between samples. This will be used to compare chromatograms from the same run sequence as well as chromatograms from different sources. The maximum number of overlays will vary from data system to data system but a minimum of 6–8 is reasonable and practicable. More overlays may be technically possible, but the amount of useful information obtained may be limited. Overlays that can be offset by an amount determined by the user are more often useful to highlight certain peak information. Ideally, the overlay screen should have hidden lines removed and be able to be printed.

2.1.9 System Suitability Test Calculations

System suitability samples are then calculated and used to determine if the analytical run meets the predetermined suitability criteria. There are various parameters that can be used to determine the suitability of a method such as:

- retention time
- signal-to-noise ratio
- theoretical plates
- resolution between two identified peaks (note that there are different resolution equations used in the United States Pharmacopeia (USP) $<621>$[8] and the European Pharmacopoeia Section 2.2.46[9])
- peak tailing and/or asymmetry

The CDS may be set up to calculate the results as the system suitability test (SST) samples are injected and commit the remainder of the samples for analysis if they are within acceptable. However, if the results are outside of the acceptance criteria, then the CDS can stop the samples from being injected (this requires the CDS to control the chromatograph).

2.1.10 Calibration

Calibration is a weak area with most data systems, as most chromatographers use many ways to calibrate their methods as evidenced by the multitude of calibration options available. Often these methods are basic and lack statistical rigour, as the understanding of many chromatographers is poor.

The main calibration method types are[9]:

- *External standard method*. The concentration of the component is determined by comparing the peak response in the sample with the response from the analyte in a reference solution.
- *Internal standard method*. Equal amounts of a component that is resolved from the substance to be determined and does not react with it (internal standard) is added to the test solution and a reference solution. The concentration

of the analyte is determined by comparing the ratio of the peak areas (heights) of the analyte to internal standard in the sample *versus* the reference solution.

• *Normalisation procedure (normalisation).* The percentage content of one or more components of the sample is calculated by determining the area of the peaks as a percentage of the total areas of all the peaks, excluding those due to solvents or any added reagents and those below the limit of detection.

Guidance for the appropriate type of calibration method to be applied to a specific method of determination is presented in Section 2.2.46 of the Pharm. Eur.[9] and is outside the scope of this book.

There are a number of calibration model options that are available in the majority of CDS systems that could be used within any chromatography laboratory. The main ones are:

• bracketed standards at one concentration or amount
• bracketed standards at two concentration levels
• response function
• average by amount
• multi-level or linear regression
• linear regression calibration curves with and without weighting
• non-linear calibration method, *e.g.* quadratic calibration

Within each calibration type, the data system must be able to cope, sufficiently flexibly, with variations in numbers of standards used in a sequence and types of standard bracketing. The incorporation of a zero concentration standard into the calibration curve should always be an option.

Each plot of an analyte in a multi-level or linear regression calibration model must contain an identifier for that calibration line and the analyte to be determined. The calibration curve should show all calibrating standards run in any particular assay. In assays containing more than one analyte, it will be necessary to interpret all the calibration graphs before the calculation of results. Again, this is an area that may be poor for data system as many only offer one line fitting method for all analytes in the run resulting in compromises.

2.1.11 User-Defined Analytical Run Parameters

The system should be capable of collating user-defined parameters (*e.g.* height, area, ratios, concentrations, *etc.*) for selected analytes from a sequence of runs. After collation system-defined and/or user-defined statistical calculations will be carried out on the data generated. The type of calculations required should include mean, standard deviation, analysis of variance and possibly significance testing.

These calculations may be unique to a laboratory but all must be documented as discussed in Chapter 7.

2.1.12 Reports and Collation of Results

Ideally, the report following an individual chromatogram should contain both elements that are user definable and those which are standard. This should enable the laboratory to customise a report. At the end of the analytical run, a user-defined summary report containing information such as system suitability results, sample ID, area or height, baseline and calculated analyte concentration should be created. This report can either be printed out or transferred to a LIMS for further analysis and interpretation.

2.1.13 Interpretation and Treatment of Chromatographic Results

One of the GMP issues since the Barr Judgement in 1993[10,11] was that outlier tests could not be used for chemical analysis data unless defined in the USP. USP 28[8] has a new chapter < 1010 > on Analytical Data – Interpretation and Treatment that can be used by non-statistical chromatographers to analyse their results and data.

The subject and discussion of chapter < 1010 > is outside the scope of this book with the sole exception that if such methods are incorporated into a CDS or data from a CDS are analysed by an external software application, it may become part of the overall system and hence covered by the CDS validation. The validation team will need to incorporate the functions and the interfaces as necessary into the validation documentation for the system.

2.2 Architecture of a Networked CDS

The scope of a typical networked CDS will consist of several hardware components as shown in Figure 7:

- *Chromatograph*. This is the instrument that performs the analytical separation and can be a high performance liquid chromatograph (HPLC), gas chromatograph (GC) or a capillary electrophoresis (CE) instrument.
- *Data acquisition*. Usually *via* an analogue to digital converter from the instrument detector to the CDS which converts the continuous analogue signal to a number of discrete digital data readings. Often the A/D unit has buffering capability if the network is temporarily unavailable and to prevent data loss. Data acquisition *via* direct digital link (*e.g.* LAN) is also an option.
- *Network*. Transport medium for moving the data from the instrument to a server for secure data storage.

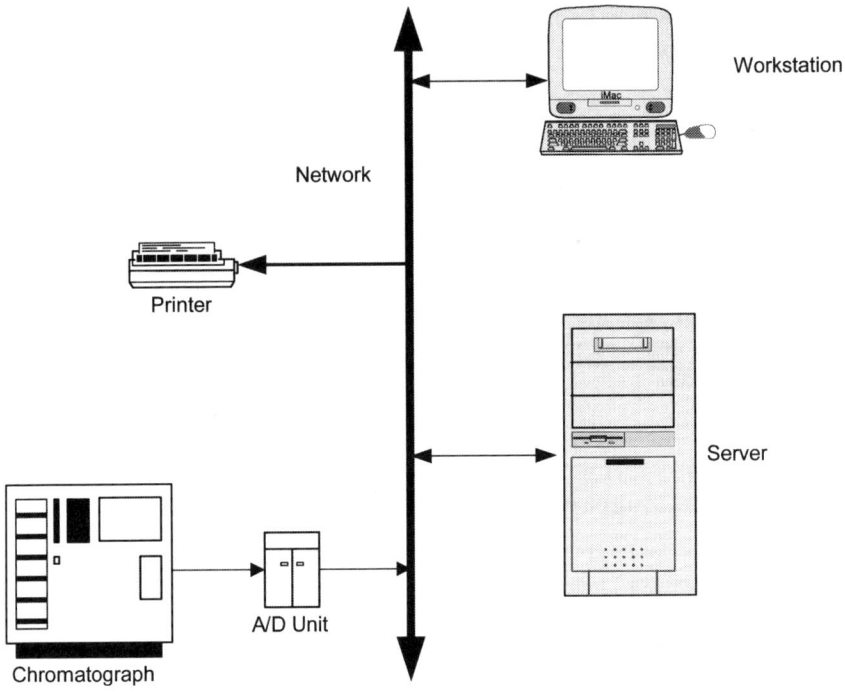

Figure 7 *Schematic diagram of a networked chromatography data system*

- *Workstation* (*client*). For operating the CDS, setting up an instrument, checking that the separation is working correctly, interpreting the resultant chromatograms after the run is finished and reporting the results.

A data system such as shown in Figure 7 can operate in a single laboratory, across a number of buildings, a whole site or between sites. The number of users and instruments can vary from tens to hundreds. The key to successful and cost-effective validation of these larger systems is the validation strategy that will be described in Chapter 8.

CHAPTER 3

Regulatory Requirements for CDS Validation

The responsibility for validation of any computerised system rests with the system or business process owner. However, most of these individuals do not fully understand the regulations they work under or the risk mitigation strategies that need to be undertaken when validating a chromatography data system (CDS). Therefore, before discussing how to validate a CDS, it is important to understand the regulatory requirements and their interpretation so that sufficient work is done to demonstrate fitness for its intended purpose and no more.

The regulations and guidelines have a view on what is expected during the implementation and release of a CDS as well as over the whole life cycle of the system. In general, the emphasis is concerned with generating the documented evidence to demonstrate that the computerised system is reliable and fit for its purpose when validated and continues to be so when it is operational and that there is sufficient proof of management awareness and control. To obtain evidence of an action usually means that it must be documented, although the format and nature of the documented evidence (paper or electronic) is left open by all schemes.

A complicating factor is the current approach of the United States Food and Drug Administration (FDA), who are currently reviewing its approach to the Good Manufacturing Practice (GMP) regulations for pharmaceutical products (21 CFR 211).[12] One strand of this approach is the development of a risk-based approach to compliance and another is the review of the electronic records and electronic signatures final rule (21 CFR 11).[13]

3.1 Regulations and Guidelines Impacting a CDS

This section will give an overview of the regulations and guidelines that a CDS will have to meet in its operation in a pharmaceutical or medical device laboratory. The aim of this section is to present the main principles of control required under the main regulations and guidance documents.

More detailed requirements and guidance will be presented in each of the main chapters dealing with the validation of the CDS application so that the regulatory requirements governing the approaches suggested in this book can be clearly linked with the regulations and understood in context.

3.1.1 FDA Good Manufacturing Practice 21 CFR 211

There are three specific sections that should be focussed on within 21 CFR 211,[12] the current GMP regulations, and how these impact the design of the overall validation of any CDS.

The first is §211.63, covering equipment design, size and location, that requires the CDS to be validated:

Equipment used in the manufacture, processing, packing, or holding of a drug product shall be of appropriate design, adequate size, and suitably located to facilitate operations for its intended use and for its cleaning and maintenance.

Interpreting this for a CDS:

- *Appropriate design ... for intended use*: The system requirements need to be defined in a document, usually called the user requirements specification. To show that the appropriate design matches the installed system, it must be tested against the documented requirements.
- *Adequate size ... for intended use*: As part of the requirements specification, the size of the system for its intended use needs to be documented. "Size" can be defined in a number of ways such as number of users, the number and nature of the chromatographs being linked to the system as well as the size of the computer hardware required to support the overall operation; in the latter instance, liaison with either the vendor or the organisation's IT department.
- *Suitably located ... for intended purpose*: The location of all the components of the system needs to follow the respective manufacturer or vendor recommendations. This will either be documented in vendor documentation or in the installation plan for the system. It will be demonstrated by the successful execution of the system installation and operational qualifications, where appropriate.

The second requirement under GMP is §211.68(b) that covers automatic, mechanical, and electronic equipment and has several issues to consider when dealing with CDS and other computerised systems:

Appropriate controls shall be exercised over computer or related systems to assure that changes in master production and control records or other records are instituted only by authorized personnel.

Therefore, a change control procedure must be operational so that authorised individuals can initiate approved changes with records of the change that can be inspected. Where necessary, approved changes must be tested and validated.

Input to and output from the computer or related system of formulas or other records or data shall be checked for accuracy. The degree and frequency of input/output verification shall be based on the complexity and reliability of the computer or related system.

If you work electronically, validate once and use automatically until changes are made to the system. The approaches for doing this are outlined in Chapter 6. If you use the CDS with manual input, transcription error checking will need to be undertaken for every analytical run. Therefore, it makes sense to eliminate paper and work electronically wherever practicable.

A backup file of data entered into the computer or related system shall be maintained except where certain data, such as calculations performed in connection with laboratory analysis, are eliminated by computerization or other automated processes. In such instances a written record of the program shall be maintained along with the appropriate validation data. Hard copy or alternative systems, such as duplicates, tapes, or microfilm, designed to assure that backup data are exact and complete and that it is secure from alteration, inadvertent erasures, or loss shall be maintained.

Data need to be protected with implementation of effective backup measures (Note: the phrasing used above is from the 1970s – ensure that the methods used are current and reflect today's requirements and best IT practices).

The third is under the section dealing with Laboratory Controls, §211.160(a):

The establishment of any specifications, standards, sampling plans, test procedures, or other laboratory control mechanisms required by this subpart, including any change in such specification, standards, sampling plans, test procedures, or other laboratory control mechanisms, shall be drafted by the appropriate organizational unit and reviewed and approved by the quality control unit.

If you purchase vendor validation or qualification materials, there needs to be an approval before you use the documentation to demonstrate that it is fit for its purpose and to review and approve the documentation after the work has been carried out. Furthermore,

The requirements in this subpart shall be followed and shall be documented at the time of performance. Any deviation from the written specifications, standards, sampling plans, test procedures, or other laboratory control mechanisms shall be recorded and justified.

Deviations to a plan must also be documented. We will look at deviations from the validation plan in Chapter 8. The fourth is §211.160(b):

Laboratory controls shall include the establishment of scientifically sound and appropriate specifications, standards, sampling plans, and test procedures ...

Here, you need to think of all the tests that you have used in the performance qualification (PQ) testing to show that the CDS is fit for its purpose and if the design is scientifically sound (see Chapter 14).

3.1.2 Good Laboratory Practice: 21 CFR 58

The regulations for equipment under GLP[14] are very similar to those under GMP. §58.61 for equipment design states:

Equipment used ... shall be of appropriate design and adequate capacity to function according to the protocol and shall be suitably located for operation, inspection, cleaning and maintenance.

Apart from a few words difference, the recurring themes are adequate design, adequate capacity (not size as per GMP) and suitably located.

3.1.3 Quality System Regulation for Medical Devices: 21 CFR 820

The medical device regulations, 21 CFR 820[15] were written in the 1990s when computerisation was more extensive than in the 1970s when the GMP and GLP regulations were drafted. Therefore, references to computers and their validation are far more explicit and encompassing.

Software used in device, production of the device or implementing the quality management system (QMS) must be validated.

- *820.30(g): Design validation: Each manufacturer shall establish and maintain procedures for validating the design device ... Design validation shall ensure that devices conform to defined user needs and intended uses and shall include testing of production units under actual or simulated use conditions. Design validation shall include software validation and risk analysis, where appropriate ...*
- *820.70(i): Automated Processes: When computers or automated data processing systems are used as part of production or the quality system, the manufacturer shall validate computer software for its intended use according to an established protocol. All software changes shall be validated before approval and issuance. These validation activities will be documented.*

Note some common features:

- Defined user needs and intended use: document requirements in a specification
- Risk analysis determines how much effort is necessary in the validation
- Established protocol: written and approved plan and testing instructions
- Changes must be validated before the change is made operational
- All activities must be documented to show they have taken place

3.1.4 ICH Q7A: GMP for Active Pharmaceutical Ingredients

ICH have developed a document (Q7A) on the application of GMP for active pharmaceutical ingredients.[16] In Section 5.4 there is a specific section on computerised systems:

5.40: GMP related computerised systems should be validated. The depth and scope of validation depends on the diversity, complexity and criticality of the computerised application.

5.41: Appropriate installation qualification and operational qualification should demonstrate the suitability of the computer hardware and software to perform assigned tasks.

5.42: Commercially available software that has been qualified does not require the same level of testing. If an existing system was not validated at the time of installation, a retrospective validation could be conducted if appropriate documentation is available.

5.43: Computerised system should have sufficient controls to prevent unauthorised access or changes to data. There should be controls to prevent omissions in data (e.g. system turned off and data not captured). There should be a record to any data change made, the previous entry, who made the change and when the change was made.

5.44: Written procedures should be available for the operation and maintenance of computerised systems.

5.45: Where critical data are being entered manually, there should be an additional check on the accuracy of the entry. This can be done by a second operator or by the system itself.

5.46: Incidents related to computerised systems that could affect the quality of intermediates or APIs or the reliability of records and test results should be recorded and investigated.

5.47: Changes to the computerised system should be made according to a change procedure and should be formally authorised, documented and tested. Records should be kept of all changes including modifications made to the hardware, software and any other critical component of the system. These records should demonstrate that the system is maintained in a validated state.

5.48: If system breakdowns or failures would result in the permanent loss of records a back-up system should be provided. A means of ensuring data protection should be established for all computerised systems.

5.49: Data can be recorded by a second means in addition to the computer system.

These regulations are similar in some respects to both the FDA GMP regulations[12] as well as those from EU GMP Annex 11[9]; they are concise as well as wide ranging. However, one issue is the comment that commercially supplied software does not require the same level of end-user testing as does an in-house developed system.

3.1.5 Electronic Records and Electronic Signatures: 21 CFR 11

21 CFR 11[17] has had and will continue to have a major impact on all CDS used in all FDA regulated laboratories. This is from both a remediation perspective and also the ability to implement and use electronic signatures in place of handwritten signatures. The latter is discussed in more detail in Chapter 6.

Interpretation of 21 CFR 11 is *via* the underlying predicate rule(s) that the laboratory works under (Figure 8). All CDS systems used in development and manufacturing where records are used to register or release products are covered by 21 CFR 11. (See Chapter 19 for more information about defining the electronic records for a CDS.)

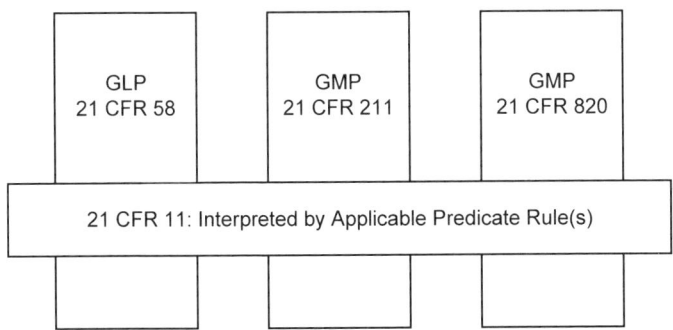

Figure 8 *Interpretation of 21 CFR 11 by existing predicate rules*

Section §11.10(a) states

Validation of the system for accuracy, reliability, consistent intended performance and the ability to discern altered or invalid records.

We need to consider a number of points for validation:

- How your validation effort links with an on-going metrology and service programme for the calibration of any analogue to digital (A/D) converters used by the system
- Can the system distinguish altered records?
- Can the system detect invalid records?
- Is the performance defined and is it reliable?

This is one of the areas that is given enforcement discretion under the FDA guidance on Part 11 Scope and Application.[18] It is important to realise that under the GMP predicate rule for laboratory records, 211.194, there is the requirement for *complete data*,[12] therefore as you change instrument control settings and review and integrate data this is a predicate rule requirement that you need to comply with.

Another important point about 21 CFR 11 is that it only refers to "systems" and never to "applications"; therefore for a networked CDS the Information Technology (IT) Department is included in the scope of regulatory compliance. Here, the IT Department needs to have procedures in place for its operations and qualify the network[19,20]; regardless of whether the IT function is outsourced or not the pharmaceutical company still retains the responsibility for the regulatory compliance.[20,21]

Some of the procedural controls (*e.g.* SOPs) required for 21 CFR 11 compliance are discussed in Chapter 15 on User Training and System Documentation.

3.1.6 FDA Withdrawn Draft Guidance Documents for 21 CFR 11

As part of its programme for 21 CFR 11, the FDA belatedly published five draft guidance for industry documents to help with the interpretation of some aspects of the regulation between 2001 and 2002. The Part 11 glossary was of little value as it added little to definitions already in the literature[22] and will not be considered further in this book.

Some selected sections of the most pertinent documents will be discussed in their relationship to CDS validation:

- 21 CFR 11 validation[23]: A System Requirements Specification (SRS) is required even for commercial software.
- Time stamps[24]: There are two points with this guidance that are directly relevant for CDS
  The first is the time stamp to be accurate to within 1 min (this would be interpreted as ±1 min) and can be either manually corrected or performed automatically using a Network Time Protocol linked to a reliable time source.

The second is the location of the time stamp as often a CDS can operate between countries and across time zones in multi-national companies. The guidance allows organisations to define where the time stamp is located. However, this must be defined in the SRS in my view and when the system is being tested it must be specifically validated and the impact of network non-availability specifically assessed as the time stamp may change if data are buffered.

- Maintenance of electronic records[25]: This topic is the most difficult part of the whole of 21 CFR 11 as the maintenance of data supporting a product registration for over 50 years may be required. This is relatively easy for paper but as we have less than 30 years' experience of managing electronic records, most of it poor, it is an unknown problem. The time capsule suggested by the FDA as a possible solution is not possible as the hardware and the skills required will not be easily available and emulation as a solution was not discussed as an option. Data migration is the only currently viable option if you still want to reprocess the data, as we will discuss later in this chapter and in more detail in Chapter 23.
- Electronic copies of electronic records[26] were issued for comment in November 2002 and the FDA state that they want electronic copies of electronic records to run on "their own software and hardware". Given the proprietary nature of CDS data files other than the actual chromatogram itself this is highly unlikely other than by inputting the method values in manually from the original system to their own one. It is perhaps for this reason that this guidance was withdrawn by the Agency in February 2003.

3.1.7 Current FDA Activities on 21 CFR 11

As part of their programme on GMP for the 21st century,[13] the FDA announced a review of 21 CFR 11 in February 2003 by publishing a draft guidance document on Part 11 Scope and Application. After receiving comments from industry, the final version of the Guidance was issued in September 2003.[18] In essence, the contents cover:

- Narrowing the guidance to cover only those records specifically required by predicate rules (these are the GMP and GLP regulations that the laboratory already operates to).
- Exempting legacy systems (defined as those in operation before 20th August 1997) from the requirements of Part 11 providing the system was operational before the effective date of Part 11 and the system met and continues to meet the requirements of the applicable predicate rules. Any changes have been controlled and validated and the laboratory can justify this.
- Enforcement discretion is given for the following sections of Part 11: validation, audit trails, copies of records and record retention.

The remainder of Part 11 is still being actively enforced, as noted in bold type in the guidance document.

The FDA is encouraging public debate about the future of the regulation and requesting feedback from industry, suppliers and other interested parties. There will be a revised 21 CFR 11 regulation that is closer to the contents of the current Scope and Application guidance issued for industry comment in 2005.

3.1.8 European Union GMP Annex 11

Some of the key sections of European Union GMP Annex 11 for computerised systems[27] are discussed below. Please note that this is a selective presentation of the clauses that are applicable for the validation and operational use of a CDS. For a full picture, please read the whole of the regulation.

Clause 11.2 states that

The extent of validation necessary will depend on a number of factors including the use to which the system is to be put, whether the validation is to be prospective or retrospective and whether or not novel elements are incorporated. Validation should be considered as part of the complete life cycle of a computer system. This cycle includes the stages of planning, specification, programming, testing, commissioning, documentation, operation, monitoring and modifying.

The regulation brings in the concept of a life cycle approach to computer validation and this will be discussed in more detail in Chapter 4.

However, the key word in this section is "novel". Where are the novel elements of the CDS application? These might be:

- Configuration of the system within the boundaries of the application. These will be specific to a laboratory and must be documented as a minimum and usually validated.
- Custom calculations performed in a specific laboratory for a specific method.
- Interfacing and control of chromatographic equipment. Control of equipment from the same vendor as the data system is under a single entity and there is a single point of contact for problems and can be relatively low risk. Where one vendor's data system controls another vendor's instrument can be very different: where is the agreement between the two companies to share the control software? If there is no agreement, the code may have been reverse engineered and you may have a problem in the future when it comes to support.

Clause 11.11 notes:

Alterations to a system or to a computer program should only be made in accordance with a defined procedure which should include provision for validating, checking, approving and implementing the change. Such an alteration should only be imple-mented with the agreement of the person responsible for the part of the system con-cerned, and the alteration should be recorded. Every significant modification should be validated.

An effective change control system is essential to demonstrate that the labo-ratory and the IT department are both in control of the CDS application and the network.

3.1.9 OECD GLP Consensus Document

OECD regulations[28] have similar approaches for maintaining the validation status of operational systems; however, this document has included the minimum requirements for written procedures. These should cover but not be limited to the following for CDS:

- Procedures for the operation and use of computerised systems (hardware/software), and the responsibilities of personnel involved.
- Procedures for security measures used to detect and prevent unauthorised access and program changes.
- Procedures and authorisation for program changes and the recording of changes.
- Procedures and authorisation for changes to equipment (hardware/software) including testing before use if appropriate.
- Procedures for the periodic testing for correct functioning of the complete system or its component parts and the recording of these tests.
- Procedures for the maintenance of computerised systems and any associated equipment.
- Procedures for software development and acceptance testing, and the recording of all acceptance testing.
- Back-up procedures for all stored data and contingency plans in the event of a breakdown.
- Procedures for the archiving and retrieval of all documents, software and computer data.
- Procedures for the monitoring and auditing of computerised systems.

It is important to realise that if you are working to GMP, you can get complementary information from GLP regulations and *vice versa*.

The requirements for a system description and change control are also well covered in this document and these will be discussed in Chapters 17 and 20 respectively.

3.1.10 FDA Guidance on General Principles of Software Validation

This document is currently the best guidance document that the FDA has written on validation of computer systems; it was issued as a draft for comment in June 1997 and released as a final version in January 2002.[29] Published by the Center for Devices and Radiological Health (CDRH) and the Center for Biologics Evaluation and Research (CBER), it provides an overview of the requirements of computer validation and explains why validation is required as software is different from hardware and many problems with software are traced to errors made in the design and development process.

The key elements that are required in this guidance document are:

- Validation is an on-going process and not a one-off event
- Plans must be established for the control and execution of the validation

- Procedures for validation must be established by the organisation
- Plans for testing are to be established early in the life cycle by documenting the intended use of the system
- Partial validation of a system is not possible

3.1.11 FDA Guidance on Computerised Systems Used in Clinical Trials

Why include a Good Clinical Practice (GCP) guidance document[30] in a chemical analysis book? The reason is that this is the only issued guidance that covers some of the 21 CFR 11 requirements for computer systems. This is important for understanding some of the basic elements of physical and logical system security and logical access control. This document covers what the FDA is looking for when they inspect computerised systems. It was reissued as a draft for comment in September 2004.

3.1.12 PIC/S Guidance for Computerised Systems

A number of regulatory agencies have formed the Pharmaceutical Inspection Convention (PIC) and this has been expanded to include EU regulatory agencies, Canada, Switzerland, Australia, New Zealand and Singapore under the banner of the Pharmaceutical Inspection Co-operation Scheme (PIC/S). This organisation has written a number of guidance documents available from the web site (www.picsheme.org).

The PIC/S guidance for computerised systems operating in GXP environments covers many of the same issues for CDS already identified in other regulations discussed above.[31] However, where the guidance is unique is where it presents six checklists to help an inspector prepare for an inspection. Reading these checklists can be useful to see where an inspector can ask to see procedures and documented evidence but also to cross-check against your validation plan and anticipated approach to see if there are any omissions.

3.1.13 Summary of Regulatory Requirements

The various regulations are very complementary; where one statute is vague, another will usually provide the information you require. In essence, the regulations ask one question: is your CDS under control?

Control is shown by having evidence of planning and being proactive:

- Validation plan for intended actions
- Specifications defining intended purpose
- The purpose of the CDS is documented and it is tested to show it works as intended
- Documents are controlled and issued before the work starts
- Testing against specifications to demonstrate fitness for purpose

- Reports are signed off in a timely manner
- The system is formally released after validation
- Changes are controlled and the system is revalidated as necessary

3.2 Warning Letters and 483 Observations Involving CDS

Some of the key 483 observations and warning letters involving CDS are discussed in this section. Under the US Freedom of Information (FOI) act inspectional observations can be requested and the warning letters from late 1996 are available for download. The companies whose non-compliances are featured here are not an all-inclusive list and the reader is encouraged to look at the FDA web site (www.fda.gov) to keep abreast of any changes in emphasis of inspections.

3.2.1 Gaines Chemical Company 483 Observations

In December 1999[32] the FDA inspected the client–server CDS operated in the QC Laboratories of the company and found the following observations:

- The CDS had never been validated and there was no documentation to assure that the system could operate as intended
- There was no change control
- No security was enabled and anyone could access the system
- No record of system configuration was made
- The application audit trail had been deliberately turned off by the staff
- No documentation of calculations performed by the system was made
- The application security could be bypassed by using Windows Explorer. This implies that files could be deleted outside of the application and with no record
- Passwords consisted of four characters and never expired
- When the system was operational anyone could access the application. The workstation had to be turned on otherwise data could not be acquired
- There were no SOPs for the operation of the system
- Backup and recovery was not demonstrated and the storage of backup tapes was not verified

These observations may reflect the situation in many small to medium-sized companies that work in the regulated environment.

3.2.2 Glenwood Warning Letter

In May 1999, Glenwood LLC received an FDA warning letter[33] that contained the following non-compliance relating to their CDS software:

Failure to validate the software programs, _____ and ____, that are used to run the laboratory HPLC equipment, during analysis of raw materials and finished products.

The _____ software does not secure data from alterations, losses, or erasures. The software allows for overwriting of original data. There are no written procedures for the use of passwords, levels of access, or data back-up.

Apart from failure to validate the CDS application, security and preservation of records were key missing items.

3.2.3 Gensia Scicor Warning Letter

A warning letter sent to the company in July 1999[34] reiterates the importance of protecting and preserving electronic records:

- *Failure to maintain laboratory records to include complete data derived from all tests necessary to assure compliance with established specifications and standards [21 CFR 211.194]. Specifically, your firm failed to properly maintain electronic files containing data secured in the course of tests from 20 HPLCS and 3 GLCS. Additionally, no investigation was conducted by your company to determine the cause of missing data and no corrective measures were implemented to prevent the reoccurrence of this event.*

The critical problem was loss of electronic records coupled with a failure to investigate the problem to stop it happening again.

3.2.4 Noramco 483 Observations

This bulk chemical company was inspected in May 2001,[35] the 483 observations highlight the detail of due diligence that any CDS validation requires in the 21 CFR 11 world. The observations are reproduced below:

- *There was no assurance that data acquired on the _____ chromatographic client–server data system was accurate, reliable and reproducible for analyses of …*
- *The CDS was not validated to ensure the system produced accurate and precise data.*
- *There was no documentation to show that the system's ability to handle overload situations in an orderly fashion.*
- *There was no assurance of the program's behaviour when working at its limit. Functional testing that includes volume and stress testing was not conducted to demonstrate the system's behaviour.*
- *Confidential and unique user logins and passwords were not assigned to each analyst to ensure data authenticity and integrity. Each workstation had a single login name and password, which was shared with all users.*
- *There were no automatic computer generated time-stamped audit trails to ensure authenticity and integrity of analytical data that was acquired and processed with the CDS. Analyst's transactions were not documented to show whether the analytical data were modified, copied or deleted.*
- *There was no documented evidence that the CDS was adequately configured and performed as intended.*
- *The firm did not have a system administrator that was responsible for system configuration and control of access to configuration tools that can modify or delete electronic records. System administrator permissions and rights were given to some QC analysts who were also responsible for analysing samples.*
- *There was no control over how analysts interacted with analytical data on the system.*
- *The universal login and password system gives users rights and permissions to edit, modify and delete data files. The system was not configured to deny analysts rights to*

directories and users did not have read/write access to analytical data on the system. Users could not only modify their records but all records on the server. There was no written documentation that established what limits and rights the IT groups assigned QC laboratory users.

- *There was no documented evidence to show that the firm periodically restored analytical data from its tape backup medium to ensure that data files could be reconstructed and were not corrupted. IT personnel did not know how to reconstruct the graphic data on workstations and referred us to analysts in the laboratory to perform system administrator tasks.*

- *There was no documentation to show that analytical data on the chromatography network could not be altered or modified by authorised users of the corporate network. The networks are connected by a router, which enables data packets to move between networks. The chromatography network did not have capabilities for tracking and controlling the integrity of each sample throughout its retention period. There were no protocols that explained the logical security procedures in place to prevent unlimited and unauthorised access to chromatographic data files.*

3.2.5 Cordis Warning Letter

In 2004, Cordis a company involved in the manufacture of drug-coated heart stents, received a warning letter[36] that contained the following observations concerning a networked CDS in operation at two of their facilities:

- *Warren, New Jersey: The automated [redacted] data acquisition system, used to ensure the integrity of the analytical data generated from laboratory chromatography equipment, was not adequately validated for its intended use. The validation did not include testing and verification of backup and restoration of the electronic data files.*

- *San German, Puerto Rico: Your firm failed to evaluate the need for revalidation of the QC Lab Data Acquisition System (which performs instrumentation control, data acquisition, data processing and report generation for all the activities performed at the San German QC laboratory) after the addition of [redacted] new acquisition servers, [redacted] new chromatographic systems and changes in the acquisition server configuration. You continued to utilize this revised QC Lab data acquisition system without ensuring that the system would perform as intended.*

In addition, prior to the approval of the Data Acquisition System Formula Validation, protocol [redacted] your firm relied on the not yet validated system for automated calculations, obtained by using custom-made formula fields, in making release decisions without manual verification.

3.2.6 Key Inspection Learning Points

Some of the key learning points from these inspections and warning letters that we need to remember for the validation of any CDS are:

- The CDS is validated and the work includes documenting any custom calculations and configuration of the system.
- Include in the PQ testing, capacity tests for stress and overload conditions to comply with §211.63 ("adequate size"). The nature and extent of these capacity tests will vary depending on the architecture of the individual CDS system and also how an individual laboratory uses it.

- Effective preservation of electronic records is vital for passing any inspection: have a procedure, follow it and have documented evidence that it works. Use redundant hardware such as RAID disks (Redundant Array of Inexpensive Disks) and uninterruptible power supplies (UPS) as a first line of defence against electronic record loss.
- Change control is vital and the process must include the IT department and the network.
- Security must be enabled, documented and tested.

The definition of Performance Qualification is "documented verification that the computer related system performs it functions in accordance with the computerised system specification while operating in its normal operating environment".[37]

The major point to make is that the laboratory must test the CDS as they use it and not how the vendor has tested it (*i.e.* in the laboratory's operating environment, using the laboratory's analytical methods, specifications and capacities and using the laboratory's networks). A vendor's IQ and OQ is not sufficient evidence that a system is fit for purpose.

CHAPTER 4

Concepts of Computer Validation

There are a number of concepts and terms that we need to get right before we start the detailed journey into validation of your CDS in the following chapters of this book.

4.1 Why Bother to Validate Your Software?

Let us start from the beginning and ask this fundamental question as there are a number of reasons for validating your CDS.

1. *Investment protection.* How much money does your laboratory waste buying software and systems that fail to meet expectations? Validation is a way of building quality into a system and increases the chances that the chromatograph and its software will meet expectations. Therefore, the investment that an organisation makes is protected from purchase on a whim or, worse, from the end of year budget spend. You know the scenario; your boss puts his/her head round the door and asks if you can spend £/€/$100,000 in 3 weeks (get three competitive quotes, assess the systems, raise the PO and have the empty box delivered to stores by the end of the financial year). Perhaps this always happens only in other organisations?

2. *Consistent product quality.* Product quality should be considered in a broad context. The product of a laboratory is information on which to base decisions. Therefore, in R&D laboratories software validation is used to ensure that the results you generate to support product development are correct. Chromatography is a key analytical technique that is heavily involved with R&D and manufacturing and it is important to know data used to develop or release a product or accept raw materials are also correct and can help to ensure consistent quality of final product.

3. *Compliance with regulations.* Both the FDA[29] and the European Union[27] expect that manual and computerised systems to show equivalent quality. Good validation practices will ease or expedite regulatory inspections and audits and hence reduce the risk of non-compliance. Confidence in computerised data enables a good foundation for management control, especially throughout a

multinational company, that can be evidenced with better communication across teams and with regulators.

4. *Protection of intellectual property.* The CDS may be used to generate and maintain intellectual property involved in the development and/or manufacture of products. These data must be scientifically correct and validation will ensure this.

4.2 What is Computerised System Validation?

Process validation was defined by the FDA in 1987 as:

Establishing documented evidence which provides a high degree of assurance that a specific process will consistently produce a product meeting its predetermined specification and quality attributes.[38]

This definition was modified by the PDA for computerised system validation (CSV) in 1995:

Establishing documented evidence which provides a high degree of assurance that a specific computer-related system will consistently produce a product meeting its predetermined specifications.[37]

The key concepts in the last definition above are:

- *Documented evidence.* There needs to be tangible deliverables to demonstrate that planned work actually took place.
- *High degree of assurance.* Software will always contain bugs and features, but in the way you use the system does it give predicable results?
- *Predetermined specification.* Without a user requirements specification the system cannot be validated.

Therefore, if there is not a well-defined specification of user requirements, and sufficient evidence of appropriate testing against these requirements, the high degree of assurance cannot be given that a given CDS system can meet its intended use with adequate capacity or size.

There are other regulatory or quality guidelines from the European Union[27] and the Organisation for Economic Development.[28] Each regulation may have slightly different requirements but all have the same basic common requirements. In general, validation is concerned with generating the evidence to demonstrate that the system is fit for the purpose you use it for and continues to be so when it is operational. In addition, there should be sufficient evidence of management control. This usually means that an action must be documented and authorised. Another feature of validation is to produce an auditable system together with the appropriate documentation to aid any audit or inspection.

The problem is how to respond to the requirement for computer validation. Any response should:

- be scientifically sound
- be structured

- provide adequate compliance
- reflect the way you use the application

This latter point is most important; *there is no point validating a function of a system that is not used.* Equally important is the fact that one laboratory's use of CDS software can be markedly different from another laboratory's use of the same software.

Computer validation must provide confidence in the system first and foremost to laboratory management and the users, secondly to an internal quality audit and thirdly to an external inspector. Inspectors only audit the laboratory on a periodic basis. All others work in the laboratory and use its computerised systems daily. The users must have the confidence in a system above all others; otherwise, your CDS investment will be wasted.

4.3 What is a Computerised System?

All CDS used in any regulated laboratory are classified as a computerised system. The key components are shown in Figure 9.[37] It is important to realise early in your project that if you are validating a computerised system, you do not just concentrate on the computer hardware and software. Validation encompasses more, as we will discuss now.

The elements comprising a computerised system consist of a computer system and controlled function working within its operating environment.

The *computer system* consists of:

- *Hardware*. The elements that comprise this part of a computerised system are the computer platform that the CDS application software runs on such as workstation or server plus clients, *etc.,* any network components such as hubs, routers, cables, switches and bridges. The system may run on a specific segment of the whole of a network and may have peripheral devices such as printers and plotters with the associated connecting cables.
- *Software*. This comprises several layers such as: operating systems of the workstation clients and networked server; network operating system in the switches and routers of the network; general business applications such as Word and Excel; CDS application software and the associated utility software such as a database or reporting language.

The *controlled function* comprises:

- *Equipment* linked to the computerised system is the chromatograph itself. The interface to the CDS can vary from a simple transmission of detector output to more complex 3D spectral acquisition. Ideally, the equipment connected to the data system should be qualified as part of the overall validation of the software; otherwise, how do you know that you are generating quality results?
- *Written procedures*. Trained staff should follow written SOPs as well as the manuals to operate the equipment and the data system software.

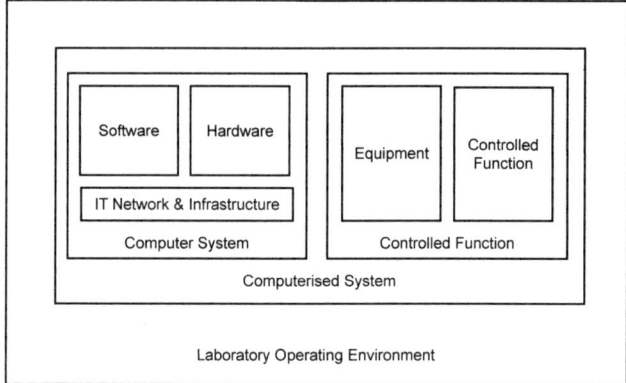

Figure 9   *Elements of a computerised system*

   Validation is not just a matter of testing system software, calibrating or testing the CDS software and any analogue to digital converters. There is a greater range of items to consider under the scope of validation. You could be open to regulatory action if you only qualify your instrument and do not validate the CDS software.

4.4   What Computer Validation is and is Not

4.4.1   Principles of Computer Validation

There are a number of principles that should be followed correctly during validation. Table 1 outlines and discusses the main ones and these are intended to reflect practical issues that have arisen when validating computerised systems.

4.4.2   Computer Validation Assumptions and Misconceptions

Many people are familiar with validation in general terms and therefore a range of assumptions exists about it, but many of these are incorrect. To help avoid these misconceptions, some of the more frequent ones are addressed in Table 2.

4.4.3   Problems with Computer Validation

There are a number of problems with computer validation:

- *Self-regulation.* Regulatory agencies take the view that the end-users of CDS software are responsible for its validation. However, the detail provided in the regulations is usually vague and it is left to the users to debate with the inspector if there is an issue. Industry guidance documents, *e.g.* GAMP,[1] is an approach but ultimately the interpretation of the regulations and the extent of validation are left to an individual company.
- *What am I to do?* This leads to the problem of how to interpret the guidelines in a cost-effective approach to validation. Often many iterations of trial and

Table 1 *Principles of computer validation*

Risk assessment	Does the system have to be validated is a key consideration at the start of any computerised system project. If the system is to be used to generate regulatory data, then validation is required, however, if it is used for research purposes only then validation is not necessary. Undertake a formal risk analysis and document the result. Further risk analysis can allow the testing to focus on key areas of the system.
Team approach	Validation generally requires support from various functions and levels within the organisation, *e.g.* scientists involved in using a system, system owner, quality assurance and if the system is networked, the IT department staff responsible for maintaining the server, *etc.* All roles involved in validation must take responsibility for their part of the validation effort.
Validation plan	There must be a formal and approved validation plan for each system. This needs to be written as early in the project as possible to avoid additional costs by writing documentation retrospectively.
Document activities	All activities must be recorded in reviewable documents which can be in either paper or electronic formats. Note carefully that it is not enough to observe the result of an activity or test and not record it. The politically correct term for this approach is "informally documented". This leads to regulatory observations and warning letters.
Four eyes principle	All documents should be written and reviewed by at least two people (*e.g.* two sets of eyes) to ensure that they are correct from both the technical and compliance perspectives.
Document your requirements	Your user or system requirements specification is your map through the system development life cycle and prevents you being seduced by technology or salespersons. Without this document, you cannot validate a computerised system.
Traceable requirements	All functions and components of a system must be traceable to approved user requirements specification or configuration documents. Furthermore, it must be demonstrated that these requirements are met within the implemented system.
Vendor assessment/ audit to assess software quality	Vendors must be assessed and if necessary audited. It is not adequate that another organisation has audited or assessed the vendor. This must be performed by your organisation to your standards. Furthermore, it cannot be assumed that products purchased from vendors are validated as this is the end-user's responsibility.
Predefined test results and acceptance criteria	All testing must be based on comparison of actual results to expected results within defined and approved test scripts. Furthermore, acceptance criteria must be explicitly stated, not implied, and based on sound scientific principles.
Documented operation	It must be demonstrated that operation of a system follows the system standard operating procedures. These SOPs must be followed by the users and must reflect current working practices with the system.

(*continued*)

Table 1 *Continued*

Independent approval	The person approving key validation documentation must be independent of the validation team, the users and the developers of the system. A quality assurance involvement from the beginning of the project is essential.
Organised archive	An archive for validation documentation must exist and it must be well organised. It must be possible to retrieve both physically and electronically archived documents accurately and quickly, or at worst within 24 h. This is essential to meet the requirements of GXP regulations.
Training and ongoing training	It must be demonstrated that all users, management, technical support and IT operations are trained in and are familiar with the system on an on-going basis and applicable regulations. This will require initial and on-going training for all types of users (system manager, supervisor, user and IT support staff).
Standard operating procedures	The system must be operated using documented and approved standard operating procedures. Further, and crucially, it must be possible to demonstrate that users continue to use the documented standard operating procedures over time.
Control and manage change	Formal change management and configuration management procedures must be applied to all configuration items of the system, *e.g.* hardware, application software, system software, training materials, SOPs and all documentation.
Define system access	Logical and physical access to the system, functions and the data must be clearly defined and validated. This needs to be updated regularly for compliance with 21 CFR 11 regulations if changes are made to user access or system functionality.
Maintain validation	Once validated, a system does not stay so automatically. The system owner needs to ensure that the system remains under control and approved changes need to be validated or revalidated when they occur, after the impact of the change has been assessed. Moreover, the system may need to be audited internally to ensure that the validation status has not changed.
System owner is responsible for validation	The business owner of each system is responsible for the validation of that system. Whilst others may carry out validation on behalf of the system owner, the responsibility for validation cannot be delegated.

error can be involved, where validation is either over-engineered or not sufficiently rigorous.

- *Complete testing of a system is a myth*. Unless there is a very simple system, it cannot be completely tested. This was demonstrated by the work of Boehm[39] who described the simple program flow segment shown in Figure 10. The number of conditional pathways and hence possible tests of the software, in this segment, were calculated to be 10^{20}. If one makes an absurd assumption that one test can be conceived, designed, executed and documented per second, then it will take more than three times the geological age of the Earth to

Table 2 *Misconceptions of computer validation*

We bought a validated system	False! Any vendor product implemented in a particular environment becomes a unique item, as the combination of environment, parameters, configuration, data content, interfaces, user procedures, *etc.*, are unique. Note, as stated above, the system owner is responsible for validation and this cannot be delegated.
	Certificates of "validation" from vendors or "validation kits" only apply to the portion of the system development life cycle that the vendor is responsible for. The system owner is responsible for the whole life cycle and, at best, these materials only provide a partial solution.
Partial validation of the system	You cannot partially validate a computerised system, it is either all (validated) or nothing (unvalidated). See the FDA guidance document on General Principles of Software Validation for further information.[29]
Long-term use equals validation	The fact that a system performs without problems for an extended length of time does not mean that the system is validated. To be validated a system requires documentary proof it meets predetermined validation criteria.
Method validation can be used to validate a CDS	Not even worth contemplating. Method validation can only used to assess if a specific analytical method is fit for its intended use. Method validation can only be carried out after the chromatographic equipment has been qualified and the CDS software has been validated.
Validation is a one-off activity	Validation is a journey and not a single event in a system's life cycle as changes to the system inevitably occur, for example, upgrade of application software or operating system. Therefore, on-going revalidation of a system is required until the system ceases operation.
	The data generated by the system need to be available for a batch's shelf-life plus 1 year for GMP data or 15 years after the last launch in the last country for a system operating in R&D.
Validation does not need documentation	Oh yes it does! All activities contributing to validation of a system must be proven to have taken place, *i.e.* documented either in paper or electronic means. If it is not written (approved and reviewed) it is a rumour (attributed to Ron Tetzlaff, an ex-FDA inspector).
GMP = giant mass of paper	The documentation needed to validate a system is little if any more than that required for good practice delivery of a computerised system not requiring validation. Further, references to vendor documentation can be used ideally when these references include author, title, date/release number, *etc.*
Validation equals software testing	Wrong again! Firstly, a system includes more components than just software, *e.g.* procedures, hardware, documentation and people. Secondly, other activities than testing are needed to prove a system functions as desired, *e.g.* system specifications.

(*continued*)

Table 2 *Continued*

Requirements are not needed	The definitions of validation above explicitly state that system requirements are required. In the absence of requirements: We cannot be certain which functions to specify, to meet business needs. A system cannot be validated to see if it meets these business needs.
Just a documentation exercise	It is not adequate just to document validation features retrospectively. Validation must be specified into a system first rather than testing quality in after installation. Further, it must be demonstrated that the use of the system in practice continues to meet designed in validation features, *e.g.* that standard operating procedures are being followed.
Validation is a job for IT or QA/QC	Validation is the responsibility of the users of the system, in particular, the system owner who is legally responsible for validation. You cannot delegate this responsibility except due to incompetence.
Regulatory bodies do not care about IT systems	Wrong yet again! There are increasing trends both to inspect IT systems and in the level of sophistication of such inspections; if in doubt see the warning letters and 483 observations in Chapter 3.

validate this program segment. Unfortunately, most CDS softwares are far more complex; therefore, procedures to record and fix errors are very important, as we will discuss in a later chapter.

- *Consistency of inspection.* The human element, in the form of what will pass without comment with one inspector or auditor but not another, will never completely disappear. The computer literacy of inspectors is increasing and with this will come increased scrutiny of computerised systems including the software, far more so than now. However, consistency of regulatory approach and inspection is highly desirable.

4.5 Computer Validation, Equipment Qualification and Method Validation

The purpose is to discuss the meanings of the terms CSV, equipment qualification (EQ) and analytical method validation (AMV). It is important to define the main

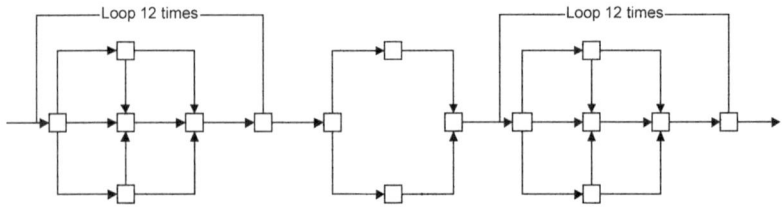

Figure 10 *Complete testing of software is impossible*[39]

terms used in this chapter so that the discussion of CDS validation is built on firm foundations. The key terms we will define here are:

- equipment calibration and adjustment
- equipment qualification
- computerised system validation
- analytical method validation.

4.5.1 Equipment Calibration and Adjustment

This is a vendor service function where according to Parriott:[40]

> *The term calibration implies that adjustments can be made to bring a system into a state of proper function. Such adjustments generally cannot be performed by analysts and are best left to trained service engineers who work for, or support, the instrument manufacturers.*

A calibration shows that a chromatograph or a module supplies you with the correct values. Depending on the numbers of sensors in your instrument, you will have to perform different calibrations. Derived from the above definition, equipment calibration can be sub-divided into:

- *Calibration*: Determination of the deviation of an instrument parameter.
- *Adjustment*: Change of an instrument parameter to meet specification.

Calibration and adjustment require reference materials or standards. These are substances with known literature values needed for calibration or traceable to recognised international standards. Calibration and adjustment are therefore inextricably linked to preventative maintenance and then to equipment qualification. Whenever calibration involves adjustments of the type described above, it is important to document the activity and, where appropriate, the user needs to requalify the chromatograph concerned. It is important to realise that this term can be confused with equipment qualification.

4.5.2 Equipment Qualification

Equipment qualification is demonstrating that an item of equipment (*e.g.* chromatograph) is fit for its purpose. This implies that all the parameters utilised by the methods run on that instrument are within tested and acceptable limits. As many methods can be specific to a single laboratory, the instrument parameters to be qualified can vary from organisation to organisation. This is an essential requirement for equipment working in a regulated environment and is the basis for all subsequent AMV work.

4.5.3 Computerised System Validation

Demonstrating that a computerised system, including any instrument control functions, is fit for its purpose is based on the PDA definition of validation presented in Section 4.2.

Most chromatographs will have an element of computer software either as firmware within the instrument or as software running on a separate PC.

4.5.4 Analytical Method Validation

Demonstrating that an analytical method is fit for its purpose. This implies that the analyst knows why the method is required and the acceptable values of key method validation parameters.

Building on these last three definitions, the differences and interactions between them will now be discussed.

4.5.5 CSV, EQ and AMV Interrelationships

The relationships between EQ, CSV and AMV are shown in Figure 11. The first three of the six blocks are the responsibility of the equipment vendor; the remaining three are the responsibility of the laboratory or end-user. The best way to describe the relationships is to use an analogy of building a house.

When building a house, you must have a firm foundation; otherwise, the structure overhead (building) will collapse. Therefore, the foundations in our analogy are to ensure that the analytical instrument or system has been designed, built and maintained correctly; this is where the instrument vendor comes in first. The instrument and any associated CDS software must be designed correctly as shown in the first building block and the vendor is responsible for constructing and testing

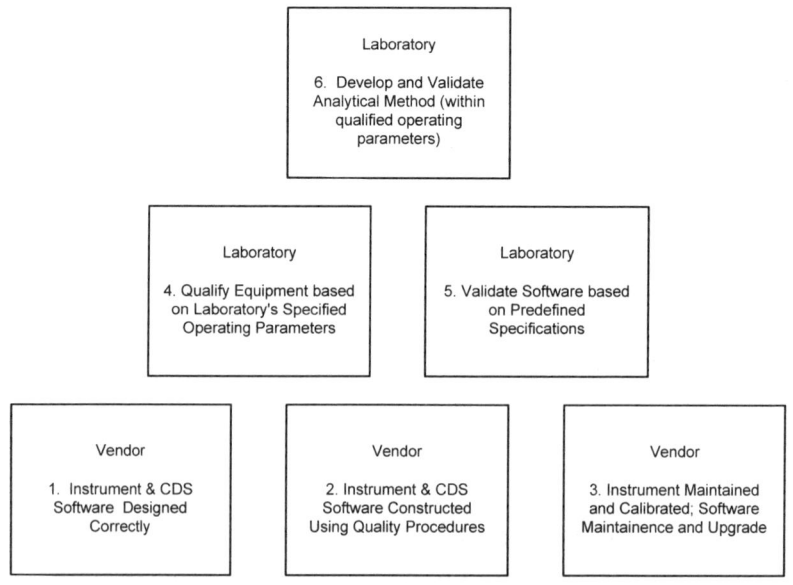

Figure 11 *Relationship between equipment qualification, computer system validation and analytical method validation*

the instrument including the software elements as shown in the second block of the foundation. Maintaining the equipment and the CDS software is also the responsibility of the vendor.

Implied, but not shown, is the role of the user who must select the right instrument and software for the right job. However, the third building part of the foundation is often missing: the laboratory must have selected the correct instrument and software system. This means that they have a specification for the functions that the instrument will perform along with the CDS software requirements including any issues on compliance with health authority regulations such as GMP and 21 CFR 11.

Once the foundations are successfully completed, the ground floor is now built. Here, the responsibility turns over to the laboratory for equipment qualification and computer validation.

The chromatograph is qualified against the range of operating parameters defined in the specification by measuring the instrument's output with a measuring device such as a digital thermometer that is calibrated to national standards to confirm that the chromatograph heater (LC) or oven (GC) is correctly calibrated. Alternatively, an internationally recognised reference standard can be used to demonstrate correct instrument performance, *e.g.* holmium perchlorate solution for UV detector wavelength.

Once the chromatograph is qualified and the CDS software is validated, each analytical method needs to be validated. Here, the method parameters to be validated, depending on procedure, are:

- precision (repeatability, intermediate precision and reproducibility as appropriate to the use of the method)
- accuracy
- limits of detection and quantification (which ones are applicable?)
- purity calculations of specific analytes that the method is measuring.

4.6 Life Cycle Approach to Validation

Under all regulations and guidance that have been issued, a life cycle approach to validation of any system, including a CDS, is stipulated. The regulations do not dictate any specific life cycle to follow and this is left to an individual company to decide.

The typical model used to explain the systems development life cycle (SDLC) is the V model. This has its origins in the 1970s and therefore it reflects an old approach to developing software.[41] This is shown in Figure 12; this V model is different from the GAMP V model[1] as the model shown more accurately represents software application development rather than automated process equipment. The qualification stages are condensed into a single stage under the control of the user shown rather than presented as three distinct stages that never occur in practice with software systems.

The left-hand side of the V represents the specification and design stages of the application, the base is the programming and the right-hand side is the testing

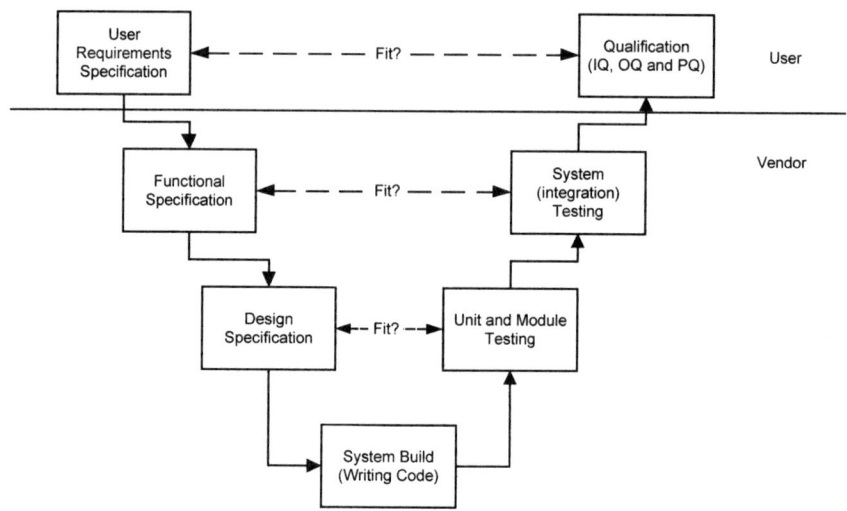

Figure 12 *A system development life cycle (SDLC) of a chromatography data system*

stages of the life cycle. It is important to realise that there is a division between the user (above the line) and the vendor (below it).

4.6.1 Roles of the User and Vendor

The horizontal line in Figure 12 indicates the separation of responsibilities between the user and the vendor of the CDS system. The user specifies, selects, installs and uses the system in contrast to the vendor who designs, builds, tests and maintains the CDS software application.

4.6.2 Design, Build and Test Phases of the V Model

The individual phases of the V model shown in Figure 12 are discussed in more detail now:

- *User requirements specification.* This specifies what the user wants the system to do. This is the basis of the user acceptance testing and qualification of the system.
- *Functional design.* This stage takes the user requirements and turns them into a computer programmer's view of the design for the system. This is an important stage that requires the crossing of disciplinary boundaries: scientist to computer programmer. From a vendor's perspective, the requirements of many laboratories are taken and incorporated into this document so that they can produce a product with as much appeal to a wide range of potential customers.
- *Design specification.* Further decomposition of the system design into individual units and modules of code. The function of each will be described, the inputs and outputs defined and the integration of all to produce the overall

system. Sometimes, this part of the life cycle may be merged with the functional requirements.

- *System build.* The actual programming of the system. This should involve programming standards to ensure that the code can be easily maintained and updated in the future.
- *Unit and module testing.* As each unit and module is completed it should be tested, first by the programmer who wrote it and then by a second independent person. As units are integrated into modules, the modules will be tested and some of the unit rests reapplied to see if functions have changed (regression testing).
- *System testing.* When all modules have been assembled into a system, there is usually a build version that is tested in-house (alpha testing). When the vendor is reasonably happy with the functions, it will be released to selected users for beta testing and feedback. When all functions are working and it is relatively bug free, the build is formally released as the next version and is available for distribution.
- *Qualification.* The new software is installed at the customer site and the users test or qualify the system to see if it is fit for its purpose. The system is then released for operational use if it is of acceptable quality.

The qualification phase for computerised systems is broken down into three phases, called IQ, OQ and PQ:

- *Installation qualification (IQ).* "Documented verification that all key aspects of the software and hardware installation adhere to appropriate codes and approved design intentions and that the recommendations of the manufacturer have been suitably considered".[1] This is the responsibility of the vendor.
- *Operational qualification (OQ).* "Documented verification that the equipment related system or sub-system performs as intended throughout representative or anticipated operating ranges".[1] Again, this is the responsibility of the vendor.
- *Performance qualification (PQ).* "Documented verification that the process and/or the total process-related system or sub-system performs as intended throughout all anticipated operating ranges".[1] This is the end-user's responsibility and can also be called end-user testing.

Qualification is presented and discussed in more detail in Chapters 13 and 14.

4.6.3 Interpreting the SDLC Deliverables for a CDS

The V model can be used to generate the documentation that could be produced during the system development life cycle. The documents that could be produced for a CDS are presented in Table 3 and the key ones are discussed in more detail in the next section. Taken together, all of these documents will provide the validation package to support the contention that the chromatography data system is fit for its purpose.

Please note that this is a suggested minimum list. You may write fewer or more documents than outlined here. The extent that an individual validation differs from

Table 3 *Validation package documentation for a CDS*

Document name	Outline function in validation
Validation plan	• Documents the intent of the validation effort throughout the whole life cycle • Defines documentation for validation package • Defines roles and responsibilities of parties involved
Project plan	• Outlines all tasks in the project • Allocates responsibilities for tasks to individuals or functional units • Several versions as progress is updated
User requirements specification (URS)	• Defines the intended purpose and functions that the CDS will undertake • Defines the scope, boundary and interfaces of the system • Defines the scope of tests for system evaluation and qualification (based on requirements)
Risk analysis and traceability matrix	• Prioritising system requirements: mandatory and desirable • Classifying requirements as either critical or non-critical • Tracing testable requirements to specific PQ test scripts
System selection report	• Outlines the systems evaluated either on paper or in-house • Summarises experience of evaluation testing • Outlines criteria for selecting chosen system
Vendor audit report and vendor quality certificates	• Defines the quality of the software from vendors perspective (certificates) • Confirms that quality procedures matches practice (audit report) • Confirms overall quality of the system before purchase
Purchase order	• From vendor quotation selects software and peripherals to be ordered • Delivery note used to confirm actual delivery against purchase order • Defines the initial configuration items of the CDS
Installation qualification (IQ)	• Installation of the components of the system by the vendor after approval • Testing of individual components • Documentation of the work carried out
Operational qualification (OQ)	• Testing of the installed system • Use of an approved vendors protocol or test scripts • Documentation of the work carried out
Configuration documentation	• Identification of chromatographs linked to specific A/D or acquisition servers • Configuration of the system operational or configuration settings • User accounts, user types and access privileges • Custom calculations, *etc.* • Technical architecture
Performance qualification (PQ) test plan	• Defines user testing on the system against the URS functions • Highlights features to test and those not to test • Outlines the assumptions, exclusions and limitations of approach

(continued)

Table 3 *Continued*

Document name	Outline function in validation
PQ test scripts	• Test script written for critical user functions defined in test plan and URS • Scripts used to collect evidence and observations as testing is carried out • Documents any changes to test procedure and if test passed or failed
Written procedures	• Procedures for users and system administrators to operate the system • Procedures written for IT-related functions • Practice must match the procedure
System description	• Overview of the operations automated by the system • Departments using the system • Linkage to other documents
User training material	• Initial material used to train super users and all users available • Refresher or advanced training documented • Training records updated accordingly
Validation summary report	• Summarises the work to demonstrate the CDS is validated • Discusses any deviations from validation plan and quality issues found • Management authorisation to use the system

this approach will depend on the amount of regulatory risk that the organisation or laboratory management wishes to carry after the validation.

4.6.4 Time Spent per Life Cycle Stage

The model, in its simplest form, illustrates a flow down the left-hand side and up the right-hand side of the model. No feedback loops are shown in any of the figures illustrating the model in this chapter. However, do not be misled that feedback loops do not exist; the number and extent of them depends on the time the user or vendor has spent on the various stages. The more time spent in the design stages means that the build and test will go more smoothly and quickly.

Rushing the URS to meet a deadline may mean that items are incorrectly specified or missed out and this may not be discovered until the test stages of the SDLC. Specification of a CDS is the inexpensive part of the life cycle for both the chromatographer and vendor. Missing or cutting short this stage means that the laboratory can select the wrong system or that required functions are not available on the system.

4.6.5 Relationships Between Phases of the SDLC

The V model is very useful for highlighting the relationships between the stages of the life cycle. Figure 12 shows these in outline. At the horizontal level there is a correspondence between the design side and the testing side. For example, the design phase is related to the corresponding unit and module test phase. The design specification will outline the individual units and modules that will need to be coded and their functions. It will also outline which individual units and modules will combine (inputs and outputs between them, *etc.*) to form the whole system. Therefore, the unit and module tests that are applied will base their test design and acceptance criteria on the specifications in the design document.

This relationship also applies to the other horizontal pairs in the model: functional specification and system test and user requirements specification and qualification phases. This last pair is important from the perspective of the users as the URS defines the tests and their limits to be carried out in the qualification or user acceptance testing.

4.6.6 User and Supplier Responsibilities

It is unlikely that you will be developing your own CDS software and therefore you will be purchasing a commercial system from a vendor. The V model is very useful for highlighting the responsibilities between the two parties. The users are responsible for the URS and the qualification or user acceptance tests, whilst the vendor is responsible for the remaining stages of the life cycle. This is illustrated in Figure 12 by the horizontal line. Above the line is the user's responsibility and below it is the vendor's.

While the V model looks good, there are some limitations to its use. As you can see, the model only covers the initial development of a system from the user specification until it becomes operational. This only represents a small fraction of the total time that a system is used. Some data systems can be operational for up to 10 or more years (including upgrades of the CDS software and the hardware platform); therefore, there should be a mechanism to accommodate this in the model.

The following life cycle phases are missing from the V model:

- *Operation.* There are a number of tasks such as backup, recovery, change control, configuration management, archive and restore that need to be covered in this part of the life cycle.
- *Maintenance.* Every time an upgrade or change is considered to the system, this part of the life cycle will be invoked.
- *Retirement.* When a system is finally retired and the data migrated to a new system or archived, there is no mechanism in the current model.

Therefore, to accommodate these later phases of the life cycle, the V model could be modified and look like Figure 13.

4.6.7 Software Implementation *versus* Development Life Cycles

The emphasis in validation literature has been on SDLCs; however, the GAMP Good Practice Guide for Laboratory Systems[42] has proposed an alternative

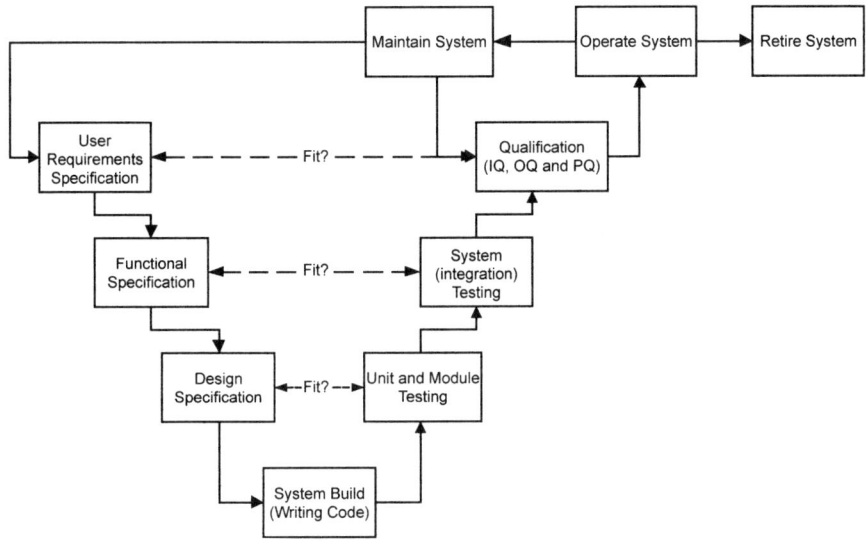

Figure 13 *Modified V model for system operation and retirement*

approach called a system implementation life cycle (SILC). The rationale for this is that most laboratory systems are commercially available and are implemented not developed. Only through a vendor audit are details on the design and development of the system available.

In essence, the SILC phases consist of:

- concept phase
- identify solutions
- request for proposal
- user requirements specification
- testing (IQ, OQ and PQ)
- system release
- system use/maintenance
- periodic review
- system retirement.

These are mapped to the remaining chapters of this book in parentheses. In essence, the SILC omits the bottom section of the V model. It aligns the V model to better reflect the commercial off-the-shelf nature of many laboratory systems. However, a CDS still follows an SDLC as it is a Category F system under this guide.[42]

4.6.8 Document Controls

Many of the documents produced in the life cycle will be controlled so this means each must have the following attributes:

- paginated correctly (*e.g.* Page X of Y)
- signed by the author

- date of signature
- document history
- effective date
- authorised by two others (technical and compliance/release reviews)
- distributed to specified individuals.

Other documents such as memoranda, notes and minutes of meetings do not need to be controlled documents. It is wise to define those documents that will be controlled in the validation plan discussed in Chapter 8.

4.7 Computer Validation Roles and Responsibilities

There are three key roles in validating a CDS from a laboratory perspective; these are the users, quality assurance and information technology (IT), where appropriate. Each role will be described below together with an outline of their responsibilities.

- *Users*. Responsible for the overall validation of the system. This is achieved by defining the system's functions, selecting the system, verifying its installation and by defining and executing the validation plan. Users will need to have standard operating procedures written for operating and supporting the application, the users must be trained and they must ensure that the complete documentation of the system is available for audit and inspection. Although the end-user is responsible for these areas, they need help, advice and support in this. Active support by management is essential for making the resources available for the validation effort and to take the responsibility for authorising the use of the system in the regulated environment. Furthermore, management must encourage the participation of the quality assurance (QA) in this process.
- *Quality assurance*. Responsible for assistance in the interpretation of regulatory guidelines for computerised systems and how they apply to the CDS. QA will review the key documentation produced during the validation effort. Monitoring of the testing and validation effort and offering assistance in developing SOPs are additional roles and responsibilities for the QA. If there are any vendor audits to be undertaken, then the QA should be involved in the planning and execution of this activity. However, some QA personnel may not be very computer literate, but this must change as many regulations involving computerised systems require the active involvement of the QA.
- *Information technology*. Responsible for help in purchase, installation and operation of the CDS software for systems running on a network. If a group is not available or the users take on this role, then the responsibilities outlined here will be transferred to the users. Responsibilities will include running the hardware and software, backups, resolving problems, *etc*. However, in offering support for a regulated system, the IT group becomes bound by the regulations or guidelines that the laboratory works under. What is not often realised both by the users and the IS group is that any unauthorised change to the operating system or network will make a validated CDS software non-compliant.

External roles may also be involved in the CDS validation project and these include the following:

- *System vendor.* The system vendor should be able to help with advice on the sizing of the system, hardware needed for good performance, assistance with vendor audits and help with qualification of the system (installation qualification and operational qualification only).
- *Validation contractor.* Writing and execution of validation work.
- *Consultants* for advice on the overall validation process or specific portions of it.

However, the organisation implementing the CDS retains responsibility for the work done by external staff.

4.8 Following the Corporate Computer Validation Policy

Many organisations have developed their own internal computer validation policies and procedures. The general approach is shown in Figure 14, where the regulations are interpreted by an organisation into a policy that simply states what should be done when validating a computer system.

The existence of a computer validation policy is not enough; underneath should be further documents that interpret the policy into more practical detail: moving from the "*what*" of the policy to the "*how*" of the interpretation. Also underneath the computer validation policy should be further guidance such as checklists and template documents for any computer validation within an organisation.

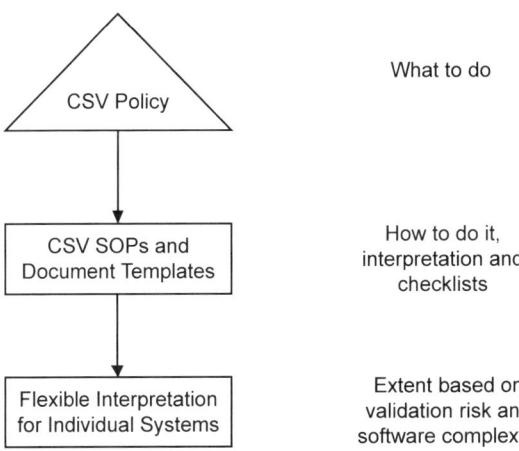

Figure 14 *Interpretation of regulations for computer systems via a validation policy for within an organisation*

The validation of your CDS needs to follow the CSV policies and procedures of your organisation. The interpretation of how much needs to be done should be left to the validation team and the system owner. However, a flexible approach needs to be taken that is based on the system risk that we will discuss in Chapter 5.

It is essential that any CDS validation follows the corporate computer validation policy and any applicable guidelines and uses available templates. If not, this should be discussed and justified in an appropriate document such as the validation plan.

CDS Validation: Managing System Risk

Risk assessment and risk management are emerging subjects within the context of computer validation. As such, this chapter covers the essence of risk management as applied to the validation of chromatography data systems.

5.1 What do the Regulators Want?

5.1.1 EU GMP Annex 11

In existence since 1992, Annex 11 Clause 2[27]:

> *The extent of validation necessary will depend on a number of factors including the use to which the system is to be put, whether the validation is to be prospective or retrospective and whether or not novel elements are incorporated.*

5.1.2 FDA Guidance on Part 11 Scope and Application

> *We recommend that you base your approach (to validate) on a justified and documented risk assessment.*[18]

5.1.3 FDA General Principles of Software Validation

Section 6.1[29]: How much validation is needed?

- *The extent of validation evidence needed for such software depends on the device manufacturer's documented intended use of that software.*
- *For example, a device manufacturer who chooses not to use all the vendor-supplied capabilities of the software only needs to validate those functions that will be used and for which the device manufacturer is dependent upon the software results as part of production or the quality system.*
- *However, high-risk applications should not be running in the same operating environment with non-validated software functions, even if those software functions are not used.*
- *Risk-mitigation techniques such as memory partitioning or other approaches to resource protection may need to be considered when high-risk applications and lower risk applications are to be used in the same operating environment.*

- *When software is upgraded or any changes are made to the software, the device manufacturer should consider how those changes may impact the "used portions" of the software and must reconfirm the validation of those portions of the software that are used. (See 21 CFR §820.70(i).)*

5.1.4 PIC/S Guidance

Section 23.7[31]

GXP critical computerised systems are those that can affect product quality and patient safety, either directly (e.g. control systems) or through the integrity of product-related information (e.g. data/information systems relating to coding, randomisation, distribution, product recalls, clinical measures, patient records, donation sources, laboratory data, etc.). This is not intended as an exhaustive list.

5.1.5 Regulatory Summary

The intent of the FDA to move towards a risk-based approach to GMP finally aligns it with the existing European approach. This allows more freedom in the overall approach to validation of a CDS but this must be justified. Risk analysis and assessment need to be documented and the decisions taken as a result approved. There are a number of phases to this. We will consider overall system risk in this chapter and later in Chapter 12 risk analysis of the individual system functions will be discussed.

5.2 Do I Need to Validate the CDS?

The first question, shown in Figure 15, considers if I should validate my CDS system.

The Computer Validation Initiative Committee (CVIC) of the Society of Quality Assurance (SQA) developed a questionnaire to determine if a computer system should be validated or not.[43] This consists of 15 questions that are phrased as closed questions (the only answer to a question is either yes or no). If you answer yes to any question, then you need to validate the system.

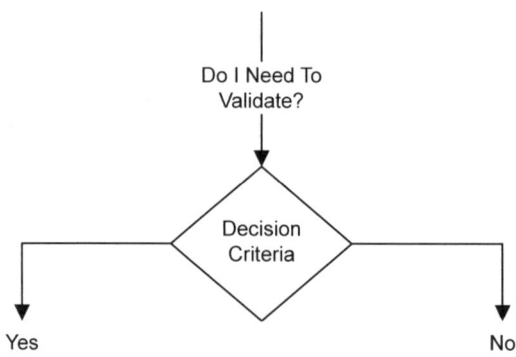

Figure 15   *Decision tree to determine whether to validate a CDS*

The CVIC questions presented below have been selected as being relevant to CDS systems:

- Does the application or system directly control, record for use, or monitor laboratory testing or clinical data?
- Does the application or system affect regulatory submission/registration?
- Does the application or system perform calculations/algorithms that will support a regulatory submission/registration?
- Is the application or system an integral part of the equipment/instrumentation/ identification used in testing, release and/or distribution of the product/ samples?
- Will data from the application or system be used to support QC product release?
- Does the application or system handle data that could impact product purity, strength, efficacy, identity, status or location?
- Does the application or system employ electronic signature capabilities and/or provide the sole record of the signature on a document subject to review by a regulatory agency?
- Is the application or system used to automate a manual QC check of data subject to review by a regulatory agency?
- Does the application or system create, update or store data prior to transferring to an existing validated system?

If you answer "yes" to any question in this list, it triggers validation of the CDS system. You should undertake this in an authorised document so that it is defendable in an inspection. Equally as important is the documenting of the "no" answers that justifies why you have not validated the system.

If you need to validate the system, this leads to the next question of "how much validation do I need to perform?"

5.3 How Much Validation do I do?

5.3.1 Balancing the Costs of Compliance and Non-compliance

There is always a question of either how much must I do or what is the minimum I can get away with when it comes to validation of computer systems, and a CDS is no exception. This can be summed as the balance between non-compliance (doing nothing and/or carrying the risk) *versus* the cost of compliance (doing the job right in the first place).

Note well the cost of compliance is always cheaper than the cost of non-compliance. If any reader is in doubt, I suggest that you read any of the recent consent decrees (*e.g.* Abbott, Schering-Plough). The cost of non-compliance can now be quantified as hundreds of millions of dollars for some companies.

Figure 16 shows the situation graphically. The vertical axes represent the cost of compliance and non-compliance, respectively. Note that the cost of compliance axis is smaller than the cost of non-compliance axis. On the bottom is the extent of compliance as a percentage.

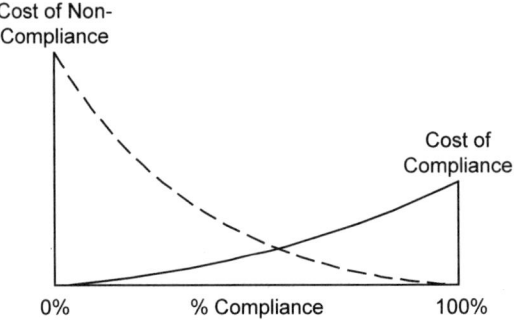

Figure 16   *Balancing the costs of compliance and non-compliance*

If all risk is to be removed, then you validate as much as possible. However, this takes much time and resource to achieve. However, if the main points are covered plus the fact that this is a commercial system, then more cost-effective validation can be accomplished in a shorter timeframe with less resource. Some risk may still exist, but it is managed risk rather than regulatory exposure.

5.3.2 Decision Criteria for Extent of Validation

The depth of validation required now depends on two major factors, as defined by the EU GMP Annex 11 clause 2[27]: the use of the system and the nature of the software itself. It is a combination of the two that will determine the extent of validation needed. The decision process is shown in Figure 17.

Some of the uses of a CDS in regulated industries are:

- active pharmaceutical ingredient (API) analysis and release
- clinical trial material analysis and release
- production analysis and release
- bioanalytical analysis from non-clinical and clinical studies

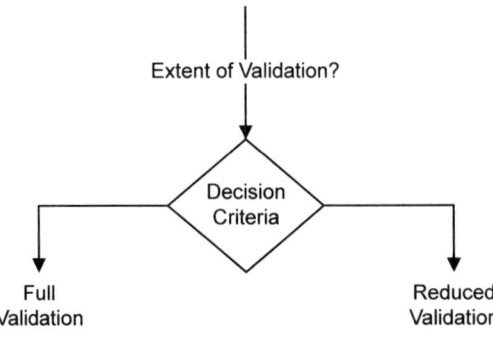

Figure 17   *Decision tree to determine the extent of validation*

There are some grey areas that can be under or outside of regulations depending how the CDS is used, *e.g. in vitro* experiments or supporting information for research data. Nature of the CDS software, *e.g.*:

- configurable application (GAMP Category 4)
- custom or novel elements of the software such as macros, custom calculations and occasionally visual basic programming (GAMP Category 5)

The higher the category, the greater the risk and therefore greater care needs to be taken to manage and mitigate the risk.

5.3.3 GAMP Software and Hardware Categories

The Good Automated Manufacturing Practice guidelines[1] in Appendix M4 provide a validation strategy for different classes of software and hardware. This concept is very important as it provides a key understanding about one of the major risk factors involved in the validation of any CDS.

5.3.4 GAMP Software Categories

There are five GAMP software categories:

- *Category 1: Operating systems.* This is the part of the software that interacts with the hardware. The validation strategy for the OS is to ensure that it is correctly installed during the IQ phase of the work and then to implicitly test the OS during the OQ and PQ phases of the qualification. The assumption being that the correct functioning of the application infers that the OS works as well. The name, version number of the OS plus service packs, is sufficient from an end-user perspective for this class of software.
- *Category 2: Firmware.* This class of software consists of read only memory (ROM) chips that are present in the chromatographic equipment and analogue to digital converters. Essentially, they are treated as equipment and are qualified and calibrated where appropriate and not validated. The sole exception is where the firmware is custom built and then this must be treated as Category 5. One issue is that the firmware version used in a chromatograph must be controlled as this can adversely impact the communication between the instrument and CDS.
- *Category 3: Commercial off-the-shelf (COTS) software.* This is a software that is commercially available and used as is from installation. The only changes that can be made are configuration of security and printers if used on a network. This is not applicable to a chromatography data system.
- *Category 4: Configurable COTS software.* This is a configurable software that is commercially available and changes to the operation of the application are made via hot buttons or switches provided by the vendor of the application (configuration or parameterisation). A CDS application software is GAMP Category 4 as it is configured: chromatographs are linked to the A/D units or

data servers, user types and their respective access controls are configured as
are the operating policies for 21 CFR 11.
- *Category 5: Custom or bespoke software.* This is the essence of "novel elements" being software that is unique to the instance or application. Typically,
  this is an application that is built or programmed in-house. However, there
  are GAMP Category 5 software elements in a CDS. These are where custom
  calculations are implemented or where visual basic code is used to program a
  function not available in the basic application.

In essence, the higher the category of software, the higher the risk of the system.
   Therefore, it is important to realise that Category 4 and 5 software can exist in
the same instance. To mitigate and manage risk, it is important to identify these
Category 5 elements in the CDS and to specify them and validate them as necessary.

5.3.5 GAMP Hardware Categories

There are two categories of hardware under GAMP 4 and these are useful to
understand them in the context of a CDS.

- *Hardware Category 1: Standard hardware.* This is a standard off-the-shelf
  hardware that is used as is with no changes. Here, the vendor's specification
  can be used.
- *Hardware Category 2: Bespoke or custom hardware.* Here, hardware is
  specifically and uniquely designed for a purpose. As such the risk is raised and
  it must be designed and tested appropriately.

Wherever possible only Category 1 hardware should be considered as otherwise
it means that more work has to be undertaken. In a CDS, Category 1 hardware can
apply to the server, workstations, data servers and A/D units used in the CDS, even
though the A/D units are uniquely designed by a supplier, they are standard
hardware as everyone who purchases this system will use them.

5.3.6 A CDS is GAMP Category 4 with Some Category 5 Software

The basic chromatography data system is GAMP Category 4 software as the system
will be configured for operation by interfacing the GC and LC chromatographic
systems (often from different vendors), specifying basic reports, defining user types
and linking the access controls to each user type plus turning 21 CFR 11 functions
on or off.
   In addition, Category 5 software can also exist in the same system where there are
macros used for automating functions, customised reports are prepared including
the situation where calculations included in the report and custom calculations are
used within the system. The rationale for this is that any CDS application cannot be
used out of the box; it needs configuration at least to acquire data from the various
chromatographs they are connected to or control these instruments.

From the perspective of this book, it is assumed that the CDS to be validated carries sufficient regulatory risk to require a full rather than a reduced validation approach. Therefore, the discussions on the life cycle and the validation of these systems will be based around this premise of a GAMP 4 software application with some elements of GAMP 5 software as implemented in an individual laboratory.

5.3.7 GAMP Best Practice Guide for Laboratory Systems

A recently published GAMP Best Practice Guide for Laboratory Systems[42] classifies a CDS as a Category F (mapped to GAMP 4) with the following elements:

- *Configuration.* Configuration parameters are stored and reused (although the configuration is proprietary, its use does not change the core software of the application).
- *Interfaces.* May have 1:1 ratio of computer to instrument interface or server to client interface.
- *Data processing.* Post-data acquisition processing done as part of the system (can analyse data with proprietary data handling algorithms).
- *Results and data storage.* Process parameters are input and stored (*e.g.* runtime and methods parameters) and the system produces raw data, test results which are stored and processed.

Therefore, this reiterates that a CDS is a GAMP software Category 4 system with GAMP 5 elements where there are custom calculations and elements.

5.4 Risk Management

ISO 14971 (Medical Devices – Application of Risk Management to Medical Devices) contains a process flow for risk management[44] that is shown in Figure 18.

5.5 Risk Analysis Methodologies

There are a number of risk analysis methodologies such as fault tree analysis, HazOp, hazard analysis and critical control points (HACCP), failure mode effect

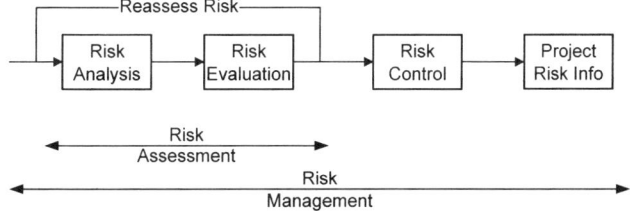

Figure 18 *The Risk Management Process from ISO 14971*[44]

analysis (FMEA) and functional risk analysis (FRA); however, only the last two approaches will be considered here.

5.5.1 Failure Mode Effect Analysis

This is the methodology described in GAMP Guide Version 4, Appendix M3[1] and suggested for use for risk assessment in computer validation. This was developed in the late 1940s for the US Military and is described in US Military Standard 1629a,[45] the approach has been further developed for the US car industry as SAE (Society of Automotive Engineers) Standard J-1739[46]; as such it is suitable for complex design and processes.

Its use in the validation of configurable COTS software (GAMP Category 4); however, is overkill as the methodology looks at assessing what can go wrong with the system, the impact of the error and if it will be picked up by the system or not. As such, the methodology is far more suitable for complex process equipment of fully Category 5 software. Therefore, we will consider a simpler methodology that can be used for a CDS that is easier to understand and apply.

5.5.2 Functional Risk Analysis

Functional risk analysis is a simpler methodology that requires a prioritised user requirements specification; the prioritisation scheme is either mandatory (M) or desirable (D). The mandatory assignment needs the requirement must be present for the system to operate and if desirable is assigned, then the requirement need not be present for operability of the system but is nice to have if it is present.[47] It is this methodology that will be described in Chapter 12.

CHAPTER 6

Process Redesign to Exploit the Tangible Benefits of Electronic Signatures with a CDS

Although not an apparent part of computer validation, process mapping and analysis is a vital first step in defining the requirements of a CDS or any computerised system. Simplification or optimisation of the laboratory process will enable a more efficient implementation of the new CDS or an upgraded application where electronic signature will be used.

The majority of laboratory processes have evolved over time; for a CDS to be effective and cost-efficient the process needs to be understood and optimised. Therefore, before starting a detailed discussion of how to validate a CDS, it is important to realise that significant business benefits can be obtained from implementing electronic signatures using a 21 CFR 11 technically compliant CDS. Thus, give serious consideration to implementing electronic signatures when implementing a new CDS or upgrading an existing system that is technically compliant with Part 11.

6.1 What do the Regulators Want?

As a general comment on the US predicate rules (GLP and GMP regulations), there are very few explicit statements where initials or signatures are required. For the purposes of interpretation, initials or signature means signature, *i.e.* an act of signing a record or records. Other statements in the regulations also require signatures such as reviewed, authorised and verified.

In a paper world, there are many other uses of initials and signatures, where this is identification of the actions of individuals and is the same as correcting a mistake made on a paper record (who did what and when). Many "signatures" are in fact the result of custom and practice rather than specific regulatory requirements.

6.1.1 FDA GMP Regulations: Number of Signatures and Order of Signing

The 21 CFR 211[12] regulations for the laboratory covers:

> *§211.194: Laboratory Records: (a) Laboratory records shall include complete data derived from all tests necessary to assure compliance with established specifications and standards, including examinations and assays, as follows:*
>
> *(2) A statement of each method used in testing the sample... The suitability of all testing methods used shall be verified under actual conditions of use.*
>
> *(7) The initials or signature of the person who performs each test and the date(s) the tests were performed.*
>
> *(8) The initials or signature of a second person showing that the original records have been reviewed for accuracy, completeness, and compliance with established standards.*

For a single chromatographic run therefore, only two signatures are required under GMP; the person who did the work and a second person who checks that acquisition, interpretation and result calculations have been performed correctly to applicable laboratory procedures. Therefore, there needs to be a technical control for the CDS software to ensure that the order of signing is correct and that the same person cannot sign in both roles. Furthermore, there is no implicit or explicit requirement for every chromatogram to be signed electronically; one signature can cover all in a single run.

6.1.2 Required Laboratory Records

The records required by GMP[12] are stated under 21 CFR 194(a). This requires:

> *Laboratory records shall include complete data derived from all tests necessary to assure compliance with established specifications and standards.*

Complete data is a rather nebulous phrase that needs to be carefully interpreted. Consideration of all actions used to generate the final results; this includes:

- All calculations factors
- Sample weights
- Calculations
- Peak integration and reintegration
- All calculations of the data generated during the run

The aim is to know how many times a sample has been interpreted and reintegrated. The key questions to be answered are "are you in control of the chromatographic methods?" and "are you fixing the results by excessive reintegration of the data?" This is linked to the FDA's aims of the prevention of adulteration and prevention of fraud.

Therefore, if a chromatographic method is registered for a specific product and the active component is being analysed; if the active peak needs manual integration, one could argue that the method is out of control. In contrast, for measurement of impurities, especially near the limits of detection, even for a registered product, manual integration is acceptable providing there is an SOP stating when manual integration is acceptable and an audit trail of the number of times a peak has been reintegrated is available. Be careful, if taken to extreme, this can invoke the Barr ruling and a laboratory can be accused of testing into compliance.

However, reintegration needs to be balanced; during a single review and interpretation session within a CDS, a peak baseline can be changed a number of times by a chromatographer (for example in the impurity situation described above); this is acceptable as the chromatographer is exercising their scientific judgement. The settings saved at the session end are the ones input to the audit trail. A second and separate interpretation session either in time by the same chromatographer or by a supervisor incurs new audit trail entries to show the changes made by the end of the second session.

6.1.3 EU GMP and PIC/S Guidance

Section 21 of the PIC/S Guidance document[27] discusses EU GMP in some depth and argues that there is no rationale either implicit or explicit that does not allow the use of electronic signatures and there is no explicit requirement for signatures to be "in writing". It does note that for read only systems, inspectors would expect to see less onerous controls than for ones that generate and manipulate data.

6.1.4 Regulations Summary

To implement electronic signatures effectively the underlying predicate rules have to be understood and interpreted: what records need to be signed? The number of signatures required for a chromatographic run is relatively few and the CDS needs to follow these requirements.

6.2 Islands of Automation in an Ocean of Paper

6.2.1 Current 21 CFR 11 Remediation Strategies

To ensure compliance with 21 CFR 11, systems need to be remediated or updated to make them meet the requirements. There are two options for Part 11 remediation strategies:

1. Replace one version of an application with a Part 11 technical compliant solution with no change in working practices when a technically compliant Part 11 solution is available from a vendor. As we shall see in this chapter this is a short-sighted approach and misses a strategic business opportunity.
2. Map and understand the current ways of working (process mapping) such as where are the bottlenecks and problems with the existing ways of working. Then improve and optimise the process so that the laboratory works electronically. This approach can streamline chromatographic analysis by eliminating duplicated or redundant operations eliminate some applications and hybrid systems resulting in reduced paper and computer validation burden.

For the purposes of this chapter, we will consider option 2 and see what tangible business benefits can be obtained from implementing electronic signatures with electronic ways of working.

6.2.2 Rationale for Using Electronic Signatures

The pharmaceutical industry requested a regulation allowing it to use electronic signatures in 1990 which cumulated in the publication of the 21 CFR 11 final rule in March 1997. The request for the regulation was to take advantage of new technology, improve efficiency and eliminate paper.

Like most good intentions, the original aim has been lost under a pile of electronic record compliance. However, it is important not to lose sight of the intent of the request.

6.2.3 21 CFR 11 is an Integrated Regulation

21 CFR 11, the Electronic Records; Electronic Signatures final rule,[17] is an integrated regulation. Subpart B (electronic records) has requirements for signing electronic records whilst subpart C (electronic signatures) has controls that are as important for ensuring the trustworthiness and reliability of electronic records as well as electronic signatures. Therefore, to use existing systems in a hybrid (electronic records with handwritten signatures) mode is just a temporary solution before working completely electronically. In this chapter, we will discuss the ways that the design of electronic signatures can be implemented in a CDS.

The prerequisite for this approach to succeed is the need for any software to be technically compliant with the requirements of 21 CFR 11. Therefore, it is important before embarking implementing electronic signatures that the software is assessed as being technically compliant with the laboratory's interpretation of the regulation.

6.3 Process Mapping and Analysis

A business process is a defined series of tasks which will produce a specific business outcome. In the context of a CDS, this process is broadly the analysis of samples by chromatography and the calculation of the final results. These results can be used to make a decision such as a batch of material is within its specification.

6.3.1 Importance of Understanding the Process

The majority of business processes have evolved over time and have not been specifically designed and chromatographic analysis is no exception to this statement. Therefore, the chromatographic process is likely to be paper based, with multiple hand-offs and multiple data transcriptions unless the process has been planned carefully.

The key principle is that to implement electronic signatures on an existing paper-based process is not just a matter of electronically signing the calculated results. It requires a different philosophy and also requires a good understanding of the regulations that an organisation has to be compliant with and the business processes that will use electronic signature.

It is unlikely that an organisation will benefit by implementing electronic signatures on an existing process unless it has been implemented to work electronically.

6.3.2 Map the Current Process

There are a number of process mapping methodologies that can be used, however, the maps that appear in Figures 19–23 are based on the Integrated Definition (IDEF) methodology of hierarchical decomposition. This is a formal way of describing a process and then decomposing it into a number of interrelated layers. There is more detail at a lower layer than the one above it. Typically for a CDS, the process definition and two layers of process description are usually sufficient to describe and understand a process.

From this and other information, the existing process can be redesigned to make it simpler and more efficient. This way the process can be electronic rather than paper[48] and make effective use of electronic signatures.

6.3.3 Other Benefits from Redesigning the Process

This chapter will focus on the benefits of redesigning the process around the use of electronic signatures, however, there are other benefits that will also emerge from process mapping and redesigning:

- Laboratory planning. When metrics of the process are available, this will enable the process redesign team to compare and contrast what factors contribute to a rapid turn around time (such as advance warning of sample delivery to the laboratory) to those when there is a very slow turnaround time (*e.g.* out of specification investigations). Understanding the reasons for both fast and slow turnaround will enable the process to be improved.
- Identification of the paper and electronic records generated in the process. This will allow those that are duplicated to be eliminated or reduced and those that remain to be protected; this latter group will include the electronic records generated by the CDS itself.
- Elimination of redundant signings and review processes that have evolved over time.
- Removal of duplicated activities that also have evolved as the laboratory has developed over time.
- Reorganisations and mergers tend to result in inefficient processes that can be improved.
- Poor planning and scheduling of work outside of the laboratory may be factors influencing the poor performance inside the laboratory.

6.4 Case Study Descriptions

6.4.1 Case Study 1

To illustrate this principle, the results from a case study where electronic signatures have been designed into the process will be presented and discussed. The CDS is installed in a pharmaceutical quality control laboratory where the system is used for both raw material and finished product analysis. There are approximately 50 users of the system. The current CDS version was not fully compliant with the technical requirements of 21 CFR 11 and was to be upgraded to a new compliant version of the software. Before the implementation of the new version, the current process was mapped and analysed to see if there were any opportunities for improvement and to make effective use of electronic signatures.

There is also a Laboratory Information Management System (LIMS) that is operational in some of the sections within the laboratories, however, at the moment there is a mixture of both lab notebooks (paper based) and a LIMS being used.

6.4.2 Case Study 2

Case Study 2 describes an active pharmaceutical ingredient (API) manufacturer, with approximately 230 users of the system and over 130 gas and liquid chromatographs. The existing process is essentially paper based with standalone systems supplied by four different CDS vendors. The majority of the systems are non-complaint with Part 11 and are being replaced with a single multi-user system that uses Citrix to terminal serve the application and control the instrumentation. The singlesite system is also technically compliant with the requirements of 21 CFR 11. As part of this approach, electronic signatures were implemented with the new system for the laboratory to gain a good return on their investment in the new CDS.

There is also a LIMS used in the laboratories, but there are no instruments or system currently connected to it; CDS and LIMS interfacing will be addressed in a later phase of the project.

6.5 Optimising the Workflow for Electronic Signatures – Case Study 1

6.5.1 The Current Process

The first task when considering implementing electronic signatures is to map the current process. This is relatively quick and the current laboratory high-level process is shown in Figure 19.[49] We can see that there are parallel electronic and paper activities when chromatographic analysis is undertaken. For example, when a chromatograph is set up, a paper record (Lab Book) needs to be updated and checked. When results are calculated the report and chromatograms printed out and the Lab Book updated and checked again.

It is important to analyse the current process and understand why you do and what you do. The majority of laboratory and chromatographic process are not designed but have evolved over time. Working practices are due to a number of

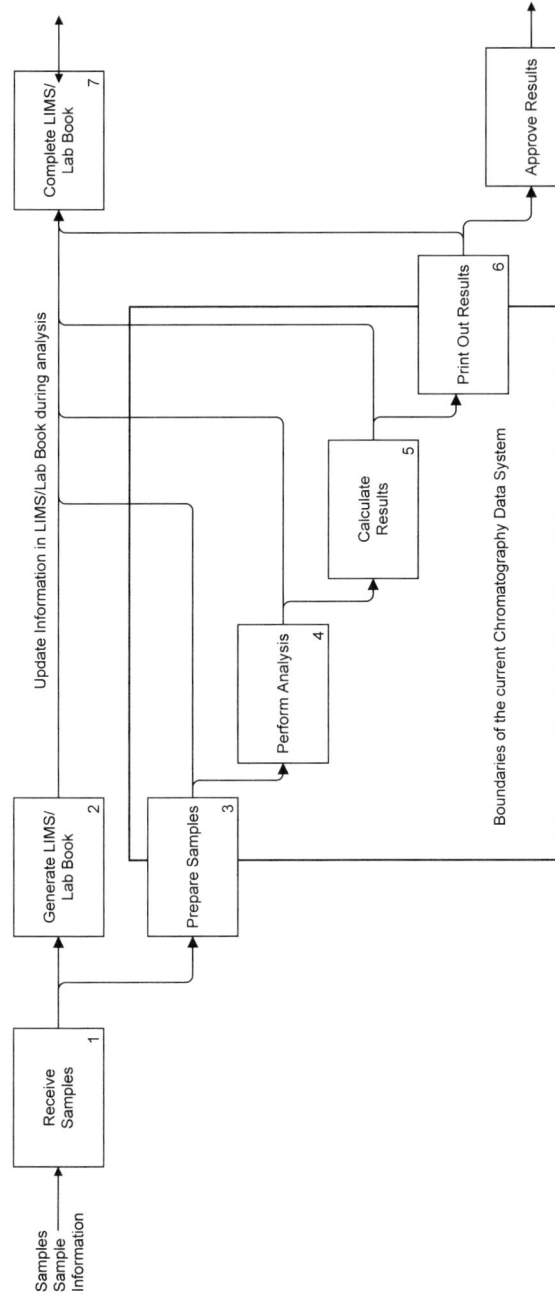

Figure 19 *The current process highlighting the boundaries of the current version of the CDS for case study 1*

factors, for example the reorganisation and amalgamation of laboratories, impact of various heads of department and response to regulatory inspections.

The general approach is that current processes have to be mapped and understood and optimised so that electronic signatures are implemented to make effective use of the CDS. Implementing electronic signatures on a paper-based process will only result in readable printed signature with no efficiency gain for the laboratory.

After mapping the process, evaluate the process metrics such as:

- How many samples are analysed?
- What are the sample turnaround times?

These data can be used to analyse the process and improve it by eliminating redundant and non-essential activities allowing it to become more efficient and effective.

6.5.2 Basic Process Improvement Ideas

Some of the main process improvement ideas for a laboratory using a CDS could be:

- Eliminating Excel and incorporating the calculations either in the CDS itself or in a LIMS.
- Interfacing the CDS to the laboratory LIMS. This allows data to be downloaded to the CDS such as sample identity, sample weight, injection sequence, and purity and salt to base conversion factors to be incorporated in the sequence file of the CDS. Then following completion of the analysis and signing of the results, the approved results can be uploaded into the LIMS.
- Reviewing of the chromatographic process to see if any tasks over-engineer quality, are duplicated or are paper based and as such can be improved under electronic ways of working.

The boundaries of the current version of the chromatography data system are also shown in Figure 19. In the current system, the approval of results occurs outside of the chromatography data system on paper.

6.5.3 The Redesigned Process

Knowing the problems and improvement ideas from the analysis of the current ways of working, a new process can be designed to exploit the use of electronic signatures. It is important at this stage to ensure that the new process is compliant with 21 CFR 11 and any predicate rule requirements and that the new version of the CDS can support the new process as well. For example, where in the process will you use signatures and where will identifications of actions be sufficient?

In the case study, the redesigned process is shown in Figure 20; the main differences are:

- Elimination of the need to update the Lab Book for chromatographic analysis. This is a quick win that is estimated to save about 0.3–2.6 FTE (Full Time Equivalents or person years). This is independent of implementing electronic signatures in the CDS.

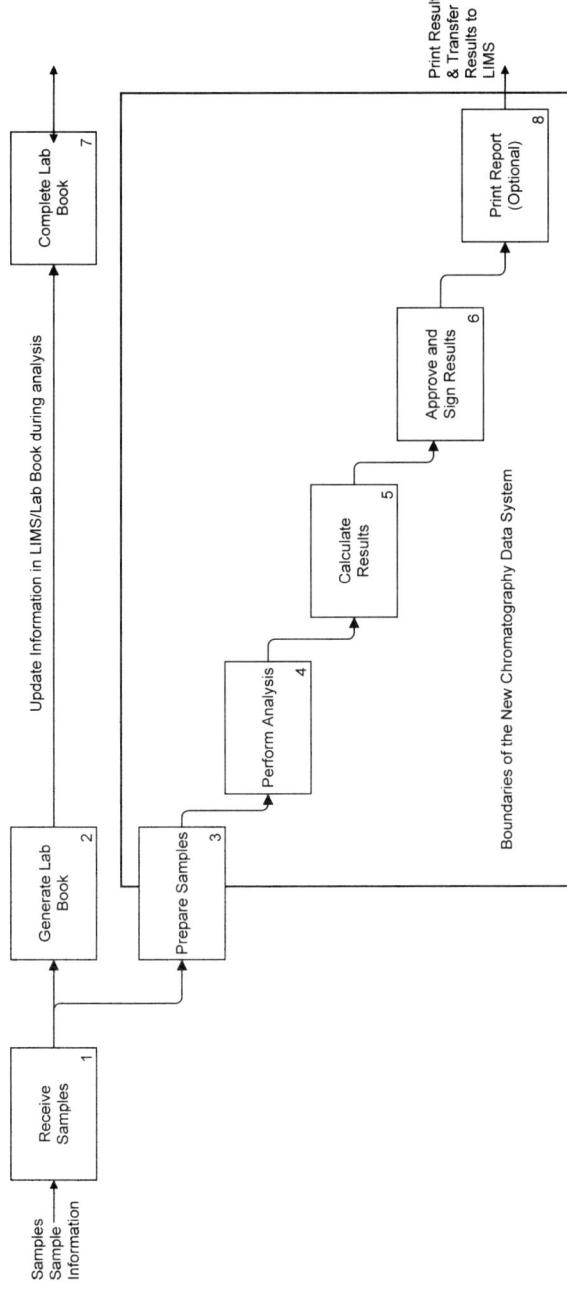

Figure 20 *The redesigned process highlighting the extended boundaries of the new version of the CDS (case study 1)*

- Expansion of the scope of the CDS. In effect the approval of electronic records and calculated results takes place in the CDS and the printout is an option.
- Configuration of the CDS to carry out all calculations rather than use a calculator or spreadsheet, this streamlines the whole process for calculating, reviewing and approving results.

The benefits of the process redesign when the CDS is linked to the LIMS would be an annual saving in the region of 6–12 FTE. This is a surprising benefit but enables more capacity to be generated with the current laboratory resources. This is against a one off cost of about 2 FTE for the process redesign, linking the system to the LIMS and validation of the CDS and the data link to LIMS.

6.6 Optimising the Workflow for Electronic Signatures – Case Study 2

Case Study 1 looked at the high-level process flow in the laboratory. In the second case study, we will look at more detail at the way that the process can be optimised for reviewing electronic records.

6.6.1 The Current Process

The high-level process for Case Study 2 is shown in Figure 21. We will focus on the approval process in task 8 in Figure 22. The approval process takes place in the LIMS and not in the CDS. Therefore, if there are any issues, the analyst or reviewer has to change systems to re-evaluate the chromatographic data.

Look also in detail at the process outlined in Figure 22, the process is very laborious and slow. Each chromatogram is signed twice, once by the analyst who generated the data and secondly by the reviewer who checks the data.

6.6.2 The Redesigned Process

When redesigning the process, it is important to have the data and the signatures in the same place. Therefore, all work associated with chromatographic data is

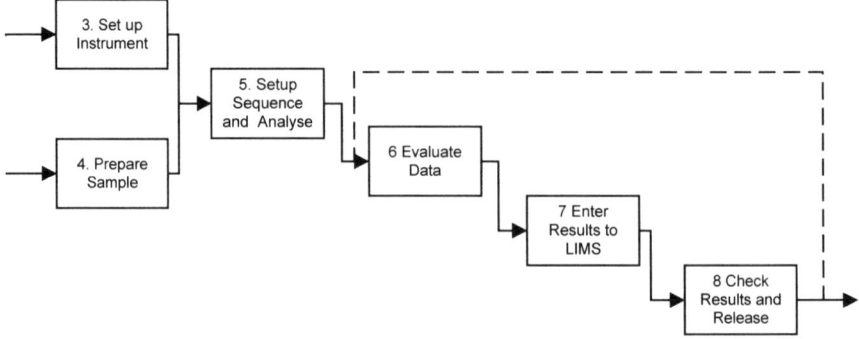

Figure 21 *Existing chromatographic process for case study 2*

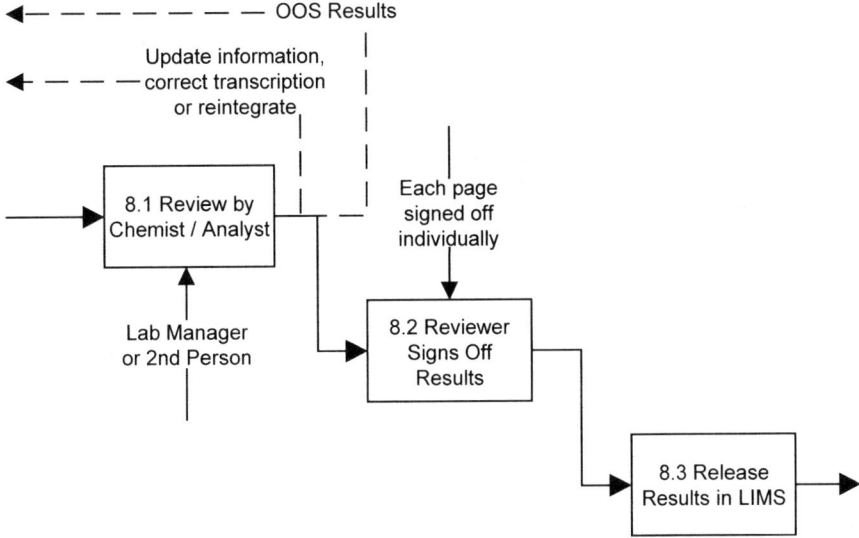

Figure 22   *The existing paper-based process for approving results for case study 2*

relocated in the CDS itself. This can be seen in Figure 23, with the subsequent electronically signed results electronically transferred to the LIMS.

Review of the results has also been streamlined and optimised as shown in Figure 24, the first and second person data review process takes place electronically within the CDS. The second person review also uses the ability of the CDS database to highlight issues that could take the reviewer time such as manual integration of individual results or audit trail entries associated with an injection. This is shown in Figure 25.

All calculations were performed with validated calculations within the system and existing Excel calculations were eliminated. Simplification of the process to eliminate paper-based tasks and work electronically within the new CDS produced savings of over 4 FTE without considering linkage to the LIMS.

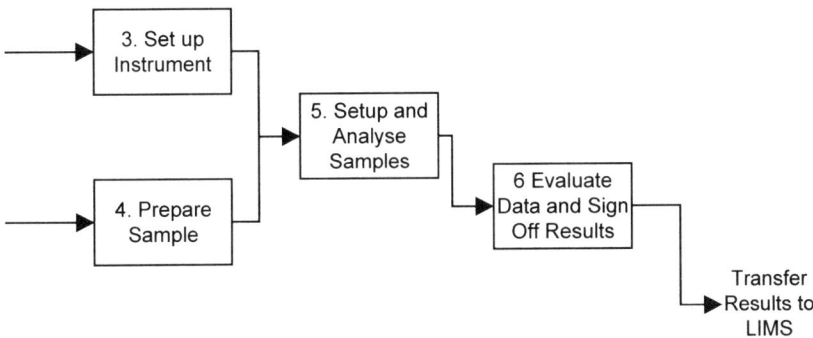

Figure 23   *Redesigned chromatographic process with electronic signatures with the CDS*

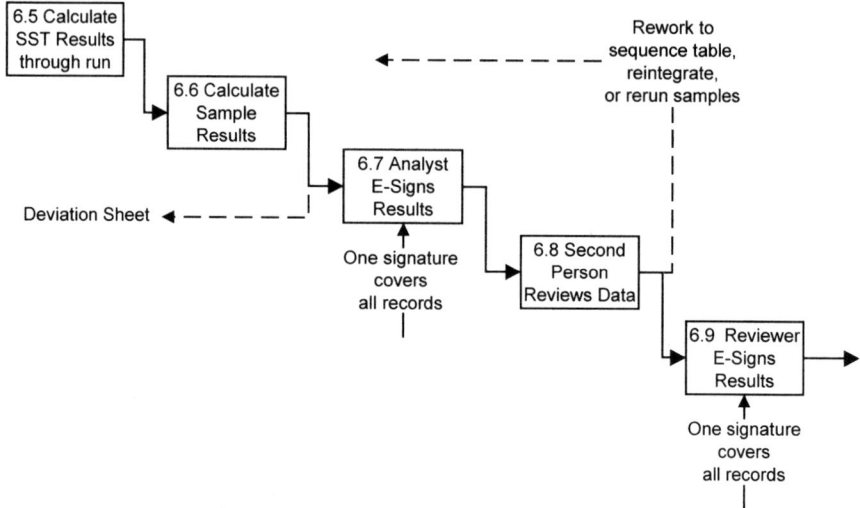

Figure 24   *Redesigned electronic review and approval process within the CDS*

6.6.3 Using the CDS for Automated Compliance

To reduce the effort needed to comply with the GXP regulations and help different users to use the CDS effectively; the system should be set up to automate regulatory compliance where ever possible.

This is shown in three screen shots in Figure 25. The first screen shot shows a review screen that has been set up to highlight to a user, where there have been issues or human intervention that could impact compliance.

The system has been set up with database view filters that illustrate if data have been modified; shown in the three screen shots of the system are three elements of this:

- *Altered*. Sample information has been changed since the chromatogram was originally acquired; sample information or weights of sample are typical information that can be changed in this way.
- *Manual*. The chromatograms have been manually integrated; this is shown in the figure where integration is noted as manual and the point that has been manually repositioned is identified.
- *Fault*. This is where an injection or run has failed a pre-established system suitability test criterion such as retention time or resolution between peaks.

Connected with the fault is the second screen that highlights the faults with the methyl phenone peak that has failed to meet the peak asymmetry SST criterion. Also shown in the second screen shot is the integration type (Int Type) that highlights if the integration is manual (bb in line 4) or automatic (BB in lines 1, 2, 3

	Sample Sets	Injections	Channels	Methods	Result Sets	**Results**	Sign Offs	Curves		
	SampleName	Vial	Altered	Manual	Faults	Result Comments				
1	PQ mix Test	26	☑	☑	☑	Manual integration of unknowns required as the chromatography was poor				
2	PQ mix Test	26	☑	☐	☑	Processed a Sample Set using Processing Method				
3	PQ mix Test	26	☑	☐	☐	Processed a Sample Set using Processing Method				
4	PQ mix Test	26	☑	☐	☑	Processed a Sample Set using Processing Method				
5	Blank	1	☑	☐	☑	Processed a Sample Set using Method Set Acetic Acid (1869)				
6	Blank	2	☑	☐	☑	Processed a Sample Set using Method Set Acetic Acid (1869)				
7	LOQ	3	☑	☐	☑	Processed a Sample Set using Method Set Acetic Acid (1869)				
8	SST	4	☑	☐	☐	Processed a Sample Set using Method Set Acetic Acid (1869)				
9	537A	23	☑	☐	☐	Processed a Sample Set using Method Set Acetic Acid (1869) forcing quantitation				
10	537A	24	☑	☐	☐	Processed a Sample Set using Method Set Acetic Acid (1869) forcing quantitation				
11	537A	25	☐	☐	☐	Processed a Sample Set using Method Set Acetic Acid (1869) forcing quantitation				
12	537A	26	☑	☐	☐	Processed a Sample Set using Method Set Acetic Acid (1869) forcing quantitation				

	Name	Retention Time (min)	Area (µV*sec)	% Area	Int Type	Resolution	Asym @ 4.4	EP Plate Count
1	Acetone	1.870	908939	44.75	BB		1.843	1144
2	*MethylPhenone*	3.130	434145	21.38	BB	5.21	*1.876*	2255
3	EthylPhenone	4.541	321246	15.82	BB	4.89	1.773	3338
4		6.343	199	0.01	bb			
5	PropylPhenone	6.770	366509	18.05	BB	1.37	1.697	4450

	Sample History
1	User: HLongden Date: 23/01/2003 16:00:46 Reason: Sample / Standard values entered incorrectly
2	Modified Vial(Batch_Numbers): <No Value> -> 57844/3
3	User: HLongden Date: 23/01/2003 16:02:05 Reason: Sample / Standard values entered incorrectly
4	Modified Vial(ColumnID): <No Value> -> 12345/a
5	Modified Vial(Column_Type): <No Value> -> Xterra C18 1 x 50

Figure 25   *Using the CDS to aid data review*

and 5). Here, an automatically fitted peak is in capital letters while manually integrated peaks have one or both baseline values as lower case letters.

6.7 Implementing Electronic Signatures Successfully

6.7.1 Understand the Process

The key to successful implementation of electronic signatures in the chromatography laboratory is to analyse the laboratory process. Implementing e-signatures on a paper driven process will not result in any net gain to the organisation. When the process is reviewed, remove repetitive and non-added value tasks and activities and also hybrid systems. Working electronically will enable a laboratory to achieve

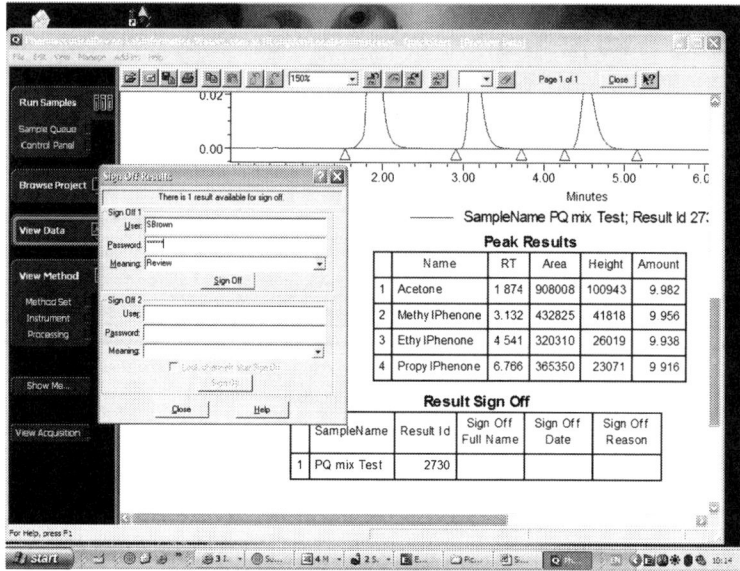

Figure 26 *Electronic signature components for a result consisting of user name, password and the meaning of the signature*

in minutes what it can take hours or days to do. The procedures required for using electronic signatures are discussed in Section 15.4.

6.7.2 Electronic Signatures Components

21 CFR 11[17] defines the minimum components of an electronic signature for a closed system as a unique combination of user identity (typically derived from the name of the user) and password (must be kept confidential and not be able to be guessed). The preamble to 21 CFR 11 allows either one signature for a single record or one signature for all records in a batch. Most CDS vendors have interpreted this regulation and implemented a single electronic signature for all records in a batch or analytical run.

Under the GMP requirements for signing (§211.194)[12] the required signatures for a batch of chromatograms will be the chromatographer who conducted the analysis and a second person who reviewed the work to confirm that it was correct and conducted to the appropriate procedures and standards. Figure 26 shows an example of an electronic signature for a chromatography data system. The components of the electronic signature (user identity and password) are input by the user, as well as selecting the meaning of the signature from a drop down menu of options that is selected by the user at the time of signing. As noted above there are a minimum of two signatures mandated by GMP, however the software can allow more signatures to be added if desired by a laboratory within the workflow of the software.

Also note that there is an option in the signature dialogue box to lock the channels (records) after sign-off by the second user. This option means that once the records are locked, another user can only view the signed records, but not reinterpret them further and so the records within the CDS match with the reported results.

Writing the User Requirements Specification

The user requirements specification (URS) is the key document in the whole of the system development life cycle that is required for both business (investment protection) and regulatory reasons (defining intended purpose). Spend sufficient time defining and writing testable requirements.

7.1 What do the Regulators Want?

7.1.1 FDA GMP and GLP Predicate Rules

Both the GLP (§58.61)[14] and GMP (§211.63)[12] regulations require that equipment be fit for intended purpose; therefore, to define intended purpose a URS is required.

7.1.2 European Union GMP

Annex 11, Clause 2[27] states:

The extent of validation necessary will depend on a number of factors including the use to which the system is to be put, whether the validation is to be prospective or retrospective and whether or not novel elements are incorporated. Validation should be considered as part of the complete life cycle of a computer system. This cycle includes the stages of planning, specification, programming, testing, commissioning, documentation, operation, monitoring and modifying.

7.1.3 FDA Draft Part 11 Validation Guidance

This withdrawn FDA draft guidance[23] discusses the main points in the life cycle of a computerised system and makes the point that a system requirements specification (SRS) is required. In the FDA's view:

Without first establishing end user needs and intended uses, we believe it is virtually impossible to confirm that the system can consistently meet them.

Put in its bluntest form: without a requirements specification you cannot validate your CDS.

> *Once you have established the end user's needs and intended uses, you should obtain evidence that the computer system implements those needs correctly and that they are traceable to system design requirements and specifications. It is important that your end user requirements specifications take into account:*
>
> - *Predicate rules,*
> - *Part 11, and*
> - *Other needs unique to your system that relate to ensuring record authenticity, integrity, signer non-repudiation, and, when appropriate, confidentiality.*

Just in case, you think that a CDS is just commercial system and you can get away with doing little or nothing, you are wrong:

> *Commercial software used in electronic recordkeeping systems subject to part 11 needs to be validated, just as programs written by end users need to be validated. We do not consider commercial marketing alone to be sufficient proof of a program's performance suitability.*
> *The end user is responsible for a program's suitability as used in the regulatory environment.*
> *However, the end user's validation approach for off-the-shelf software is somewhat different from what the developer does because the source code and development documentation are not usually available to the end user.*

7.1.4 PIC/S Guide

Section 9.2[31]:

> *When properly documented, the URS should be complete, realistic, definitive and testable. Establishment and agreement to the requirements for the software is of paramount importance. Requirements also need to define non-software (e.g. SOPs) and hardware.*

7.1.5 General Principles of Software Validation

Section 5.2.2[29]:

> *The software requirements specification document should contain a written definition of the software functions. It is not possible to validate software without predetermined and documented software requirements.*

7.1.6 Regulatory Summary

A URS is essential for the validation of any CDS operating in a regulated environment. Requirements must be testable, traceable and some requirements may indicate that a procedure needs to be written.

7.2 Business Rationale for Writing a URS

How much money does your organisation waste on buying computer systems that do not work or do not meet their initial expectations? The number of CDS systems that have no or inadequate user requirements typically outnumber the systems that have adequate specifications. A well-written URS provides several specific benefits, as it:

- Serves as a reference against which off-the-shelf commercial products are selected, evaluated in detail, and any enhancements are defined. You are less likely to be seduced by technology or buy a poor system using this approach.
- Reduces the total system effort and costs, since careful review of the document should reveal omissions, misunderstandings and/or inconsistencies in the specification and this means that they can be corrected easily before you purchase the system.
- Provides the input to user acceptance test specifications and/or qualification of the system.

A URS defines clearly and precisely, what the customer (*i.e.* you) wants the system to do, and should be understood by both the customer and the instrument vendor. The URS is a living document, and must be kept updated, *via* a change control procedure, throughout the computer system life cycle. After purchase, if and when you upgrade the software, the URS is also updated to reflect the changes and new functions in the latest version.

A URS defines the functions to be carried out, the data on which the system will operate, and the operating environment. Ideally, the emphasis is on the required functions and not the method of implementation as this focuses on the "what" rather than the "how".

If you are selecting a new CDS, then the main purpose of a URS from a business rationale is to select a system based on your laboratory's defined requirements. This avoids the user community from being seduced by technology as the system selection is based on documented requirements and what the users actually want. Furthermore, it allows the selection to be based on objective requirements that will allow you to cut through the marketing literature and focus on your specific requirements. As the URS defines what the system can do, it provides a platform to assess if any system can provide the required functionality.

7.3 Contents of a Chromatography Data System URS

7.3.1 When to Write the URS

The URS is usually the first document to be written in the life cycle validation of a CDS. The rationale for this approach is that some of the requirements specified might impact the validation strategy for the system such as a phased roll-out and this will need to be written into the validation plan.

7.3.2 Link the URS to a Specific Software Version

The URS for a CDS, or indeed any computer system, is a living document and must be linked to the specific version of the application software that is being validated;

for example, ChromSystem version 4.1. If the system is being updated to a new version, the URS must be reviewed and revised to be applicable to the new version of the software (*e.g.* ChromSystem version 5.2). Therefore, ensure that the version number of the CDS application software is written in the title and introduction of the document as a minimum.

7.3.3 Sections of the URS

From the V model shown in Figure 12 you can see that the user requirements are related to the tests carried out in the qualification phase (typically either the OQ or the PQ). Therefore, it is important to define the requirements for the basic functions of the CDS, the adequate size, 21 CFR 11 requirements and consistent intended performance in the URS that will be tested before the system goes live.

The main elements in a URS are:

- Corporate requirements for hardware, workstations and operating systems, *e.g.* terminal emulation such as Citrix.
- Overall system requirements such as number of users, locations where the system will be used and the instruments connected to the system.
- Compliance requirements from the predicate rule and 21 CFR 11 such as:
  Open or closed system definition
  Security and access configuration of the software application including user types
  Data integrity
  Time and date stamp requirements
  Electronic signature requirements
- Defined data system functions, which should be based on the CDS workflow outlined as in Figure 5 is the best framework for writing a URS. Therefore, if you have mapped the process (Chapter 6), this makes an ideal reference and prompt for formulating the requirements as they can be defined against each activity in the process. In addition, include requirements for system capacity such as maximum number of samples to be run, custom calculations and reports for the initial implementation and roll-out, *etc.*
- IT Support requirements. These include backup and recovery, off-line archive and restore.
- Interface requirements. For example, will the CDS be a standalone system or will it interface with a LIMS and if so how?

Table 4 presents a suggested list of the main sections of a URS for a CDS. This is a robust approach based on a number of validations of systems in many organisations. For a large client–server or terminal server CDS system there can be up to 500–600 requirements depending on the nature of the work that the system automated.

This idea of documenting what we want in sufficient detail sounds great, but it means more work, does not it? Yes, this is true but consider the overall benefits to you and the laboratory. The more time you spend in the specification and design phase getting your ideas and concepts right the quicker the rest of the life cycle will

Table 4 *Suggested contents of a user requirements specification for a CDS*

Section	Contents
Introduction	Purpose and scope of the document
	Referenced documents
IT requirements	Hardware specification for workstations and servers
	Operating system and database specification, *e.g.* Oracle 9i and version number
System requirements	Outline system capacity defined, *e.g.* number of users, chromatographs to be interfaced
Compliance needs	Open or closed system definition for the CDS
	Predicate rule and 21 CFR 11 requirements for e-records
	Predicate rule and 21 CFR 11 requirements for e-signatures if used
CDS functions	Operation of the system in the laboratory from set-up, instrument control, data acquisition, integration and review of the data, SST calculations and acceptance criteria, calibration models used, results calculations and reporting
Support needs	Backup and recovery of the system
	Housekeeping
	User account management
Interfaces	Is the CDS standalone or interfaced with a LIMS: if the latter then what, how and when data are transferred between the two systems
Data migration	Migration requirements from existing version to new version of same software
	Migration from old to new application with old system retirement
Appendices	Glossary
	Terms and definitions

go, as you know what you want. You will get a CDS that meets your requirements more fully and rather than find out when the system goes live that it cannot perform certain functions.

7.3.4 General Guidance for Writing the Requirements

The following guidelines should be followed during the production of the specification:

- Each requirement statement should be uniquely referenced and no longer than 250 words.
- The URS should be consistent and requirement statements should not be duplicated or contradicted.

- Specify requirements and not design solutions. The focus should be on what is required, but not how it is to be achieved.
- Each requirement should be testable. This allows the tests to be designed as soon as the URS is finalised.
- Both the customer and the vendor must understand the document. Therefore, jargon should be avoided wherever possible and key words are defined in a specific section in the document.
- Requirements should be prioritised as mandatory or desirable.
- The URS should be modifiable but changes should be under a formal control procedure.

A URS is correct if every stated requirement has only one interpretation and is met by the system. Unfortunately, this is very rare.

7.3.5 URS Issues to Consider

When defining your requirements some or all of the following will need to be included in the URS depending on your ways of working and the chromatography instrumentation that you will connect to the CDS:

- Data capture rates for all chromatographic techniques connected to the CDS will need to be specified. For example, conventional chromatography with a run time in the order of 20 min requires a data capture rate of 1 Hz. However, for capillary GC 10–20 Hz is more appropriate and for CE a higher rate still may be required depending on the overall migration time, analyte peak shape and width.
- Specify the calibration models that you will or intended to use. There are many different types of calibration model available in a CDS, however you only need to validate those models you actually propose to use; these need to be documented in the URS and all other calibration models are excluded from the validation. There is a downside in this approach, if a new calibration method is used that has not been validated then there is a regulatory risk and any new calibration method needs to be validated after an approved change control request. However, this can be managed *via* the change control process that is described in Chapter 20.
- Depending on your data system, several chromatographs may be linked into a collection workstation or an A/D unit. Here, consider if crosstalk (the interference from one channel to another) could be an issue if the A/D chip is multiplexed across two or more channels. Alternatively, consider if the total sampling capacity of the data collection and buffering unit or server is adequate for the proposed systems.
- Has the maximum number of injections for an analytical run been defined? This is a critical component, if 100 vials are routinely injected in a run, the system cannot be tested with a run of only ten samples as a user has not demonstrated adequate size or adequate capacity. The specification must match the use of the system including replicate injections.

- Some data systems will be configured to collect data from diode array detectors (DAD). If this is required, especially to analyse product, then the data collection and analysis will need to be checked as part of the adequate size as some data files can be in the Mb range. The file delete option should not be enabled to protect the electronic records generated.
- Virtually all client–server CDS systems will have a buffering capacity within their A/D or data collection units. Therefore, part of the adequate size requirements must be the ability to capture and buffer data if the network is unavailable, followed by the successful transfer of data to the server when the network connection is re-established.
- How many users will there be on the system at the same time and will the system still perform its functions reliably? This number may be lower than the number of concurrent users that you have a licence for but this is a major requirement to define in the URS and test during the PQ. If the system becomes unreliable or unstable as the number of users increases then the system owner cannot state that the system has adequate size or can perform as intended.

These are some of the considerations for each installation of a CDS. Once installed in a laboratory environment and on the organisation's network it becomes unique. The number of users, network components, server components, operating systems, software patches and laboratory configuration make it so, therefore you need to demonstrate that it works under your operating environment.

7.3.6 Making the Requirements Traceable

Although not mentioned specifically in the regulations but stressed in the various guidance documents,[1] traceability of system requirements to the testing phase is important for any system including a CDS. Therefore, the way that system requirements are presented and managed is important. It is all very well the regulations stating that a user must define their requirements in a URS, what does this mean in practice? Table 5 illustrates one way that capacity requirements can be documented; each requirement. Note that each requirement is:

- Uniquely numbered
- Written so that it can be tested, if required, in the PQ.
- Prioritised as either mandatory (M = essential for system performance) or desirable (D = nice to have and the system could be used without it). This prioritisation can be used in risk analysis of the functions and also for tracing the requirements through the rest of the life cycle as will be discussed in Chapter 12.

Remember that the URS functions are related to the tests carried out in the qualification phase of the life cycle. Therefore, if you have not specified the requirements in this document; how can you test them?

Table 5 *How system requirements for CDS capacity can be documented*

Req. No.	Data system feature specification	Priority M/D
3.3.01	The CDS has the capacity to support 10 concurrent users from an expected user base of 40 users	M
3.3.02	The CDS has the capacity to support concurrently 10 data A/D data acquisition channels from an expected 25 total number of channels	M
3.3.03	The CDS has the capacity to support concurrently 10 digital data acquisition channels from an expected 25 total number of channels	D
3.3.04	The CDS has the capacity to control concurrently 10 instruments from an expected 20 total number of connected instruments	M
3.3.05	The CDS has the capacity to simultaneously support all concurrent users, data acquisition and instrument connects whilst performing all operations such as data reprocessing and reporting without loss of performance (maximum response time is <10 s from sending the request) under peak load conditions	M
3.3.06	The CDS has the capacity to hold 70 GB of live data on the system	D

The key point for traceability is that requirements are individually numbered. Requirements traceability will be discussed further in Chapters 12, 14, and 18.

7.3.7 Reviewing the URS

Ideally, an independent group of users (persons not involved in writing the document) should evaluate the URS and challenge each requirement including any interfacing requirements for chromatographs or any other computer applications. If any missing requirements or inconsistencies can be found at this stage they are usually easy and inexpensive to correct. Therefore, the extra work in ensuring that the URS is correct are time and resources well spent. Problems that can be rectified at this stage are far cheaper to solve than those identified later in the life cycle. When the URS is complete, the outline selection tests can be generated that can be used to select a potential system and reused later in the life cycle during the PQ testing.

7.4 Writing Testable Requirements

The key to a well-written URS for a CDS, or any other computerised system, is testable requirements. However, a major problem is that many CDS users do not have an idea of how to write a user requirement and hence the quality of the overall validation effort falls at the first fence. In this section, we will look at the ways to write testable requirements.

7.4.1 How not to do it

To illustrate some of the problems of writing testable user requirements, here are two examples of how not to write requirements for a CDS. Note that both requirements are uniquely numbered which is good.

> Performance Issue 6.1.8.1: Operating at *normal* PC response times with no *undue* delay in response at *low* computer utilisation.

In requirement 6.1.8.1 there is the use of wording that makes the requirement untestable. The words normal, undue and low render the requirement useless and incapable of being tested (*e.g.* weasel words).

> Reporting 6.2.4.1: Report production at a rate of at least a page every 10 s at *modest* network and server utilisation.

Requirement 6.2.4.1 is marginally better as "report production at a rate of at least one page every 10 s" is testable and specific. However, the requirement then snatches defeat from the jaws of victory with the phrase "at modest network speed" which renders it untestable as "modest" cannot be defined. Therefore, we need to have a better approach to writing user requirements.

7.4.2 Writing Well-Formed and Testable Requirements

The recommended guide for writing software requirements is IEEE Standard 1233, entitled "A Guide to Writing Software Requirements",[50] which states that a well-defined requirement must have the following attributes:

- *Capability*. Capabilities are the fundamental requirements of the system and represent the features or functions of the system needed or desired by the user. A capability should usually be stated in such a way that it describes what the system should do and in a way that is solution independent.
- *Condition*. Conditions are measurable quantitative attributes and characteristics that are stipulated for any capability. A condition further qualifies what is required from a capability and allows the capability to be designed, evaluated or tested. This is the critical element that is usually missing from a requirement.
- *Constraint*. Constraints are requirements that are imposed on the solution by circumstance, force or compulsion, *e.g.* regulatory or corporate standards. Constraints absolutely limit the options open to the laboratory of a solution by imposing non-negotiable boundaries and limits. Examples of constraints can include interfaces to already existing systems where an interface cannot be changed, the need to change passwords regularly and the use of a specific operating system or database to meet corporate standards. Constraints can be written either as standalone requirements themselves (*e.g.* the database will be Oracle) or as constraints upon individual capabilities (as shown in the password example below).

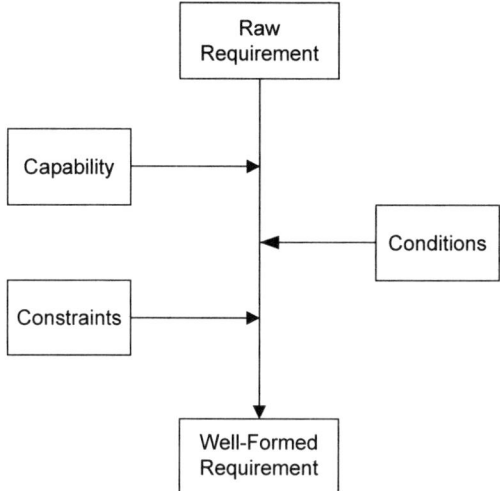

Figure 27 *Writing well-formed requirements (taken from IEEE Standard 1233)*

To illustrate the process of how to write well-formed requirements, see Figure 27.

Therefore, an example of a well-formed requirement for a CDS is the need for a password that can be used to access the system as well as be part of the component of an electronic signature. The three elements needed are shown below:

- *Capability*: the system will have alphanumeric passwords
- *Condition*: the length of which will be a minimum of six characters
- *Constraint*: and these will be changed automatically every 90 days (to meet regulatory requirements and corporate IT policy)

7.4.3 Key Criteria for User Requirements

A well-formed requirement[50] is a statement of system functionality that:

- Can be tested
- Must be met or possessed by a system to solve a user problem or to achieve a user objective
- Qualified by measurable conditions
- Bounded by constraints

Therefore, armed with this knowledge and a little practice your ability to write better requirements should improve.

7.5 Documenting System Configuration and Customisation

If you are purchasing a new system there will be not be the information available to include in the URS some of the configuration aspects of the CDS. However, if the system is already installed and you are either validating an upgrade or retrospectively validating your existing system, then you have an option to include some of the system configuration details in the URS.

Note the use of the word option. It can be included in the URS where you can add sections on the access control and user types needed along with the access privileges. In addition, the URS can contain the system configuration details, *e.g.* turning electronic signature functionality on or off as well as some custom calculations if required.

An alternative approach can be to have this information in separate configuration documents (some people may refer to these as either functional specifications or design specifications). There is no right or wrong way to do this as long as it is documented.

This part of the specification of the system must be kept up to date. If you are validating a new system these requirements may change as you gain understanding about how the system operates. However, there is a chance that if fixed early in the life cycle, they may become outdated by the time the system is rolled out.

CHAPTER 8

Controlling the Work: The Validation Plan

The validation plan provides the control of the whole of the validation effort. This is a roadmap for identifying the personnel who are involved, what they do and what documented evidence will be produced at which stage of the SDLC. The validation plan provides an organisation with the documented evidence of intent for CDS validation.

8.1 What do the Regulators Want?

8.1.1 General Principles of Software Validation

Section 4.5 of this document[29] covers Plans:

> *The software validation process is defined and controlled through the use of a plan. The software validation plan defines "what" is to be accomplished through the software validation effort. Software validation plans are a significant quality system tool. Software validation plans specify areas such as scope, approach, resources, schedules and the types and extent of activities, tasks, and work items.*

8.1.2 FDA Draft 21 CFR 11 Validation Guidance

Some of the key points from this withdrawn draft guidance document are quoted below.[23] A Validation Plan is a required document as the FDA considers that it is a

> *"strategic document that should:*
>
> - *state what is to be done,*
> - *the scope of approach,*
> - *the schedule of validation activities, and*
> - *tasks to be performed.*
>
> *The plan should also state who is responsible for performing each validation activity. The plan should be reviewed and approved by designated management".*

8.1.3 PIC/S Guidance Document

Interestingly enough, there is no mention of a validation plan in the main text of the PIC/S guidance.[31] The only reference comes in Table 6 of the Inspector's audit guide at the back of the document where it states:

> *The Validation Plan should define the activities, procedures, and responsibilities for establishing the adequacy of the system. It typically defines what Risk Assessments are to be performed.*

8.1.4 Regulatory Requirements Summary

A validation plan is the key to controlling the whole validation effort for a system: who is involved and what do they do. The validation plan is a controlled document that is approved by management.

Table 6 *Validation plan outline format (based on IEEE Standard 1012[51])*

---

1. Purpose
2. Reference documents
3. Definitions
4. Validation overview

    Roles and responsibilities

    Project plan (ideally with cross-reference to an external project plan or GANTT chart)

    Tools, techniques and methodologies
5. Life cycle validation

    Validation strategy

    Concept phase (an optional part of the life cycle)

    Requirements phase (specifying the user and technical requirements)

    Design, implementation and test phases (vendor building and testing the system)

    Installation and checkout phase (IQ, OQ and PQ phases)

    Operation and maintenance phase (change control, *etc.* during operation of the system
6. Software validation reporting

    Validation administration procedures

    Anomaly reporting and resolution

    Task iteration policy

    Deviation policy
7. Control procedures

    Standards, practices and conventions, *e.g.* corporate validation policy

    Variation and justification of variation from corporate standards
8. Validation documents

    Collated list of anticipated documents to support the validation of the system

---

Although all items listed in Section 8.1.2 are part of a validation plan, I disagree that the schedule needs to be included in the validation plan itself. Timescales invariably slip and I would recommend from a practical perspective that the schedule or project plan is maintained outside of the validation plan. This can also allow a laboratory to write the validation plan to cover the whole life cycle instead of a specific version.

8.2 What do We Call This Document?

The name for this document varies greatly from laboratory to laboratory and organisation to organisation, *e.g.*

- Validation Plan
- Master Validation Plan
- Validation Master Plan
- Quality Plan

Regardless of what it is called in an organisation, it should cover those steps to be taken to demonstrate the quality of the CDS in the laboratory. Ideally, it should be written as early in the process as possible to define the overall steps that are required and the documents to be produced from each phase of the life cycle.

As Validation Master Plan has a specific meaning under EU GMP Annex 15[27] and PIC/S[31] documents, I will only refer to this document in this book as the Validation Plan.

8.3 Content of the Validation Plan

The content of a validation plan is listed in Table 6, this is taken from IEEE Standard 1012 for validation and verification plans.[51] The purpose of the validation plan is to define the validation documentation to be produced during the initial stages of the life cycle, the roles and responsibilities involved and to provide a plan of intent for the life cycle. While the GAMP Guide provides[1] an outline for a validation plan, I have chosen to give the validation plan outline from the IEEE standard as there is much more detail and advice about what goes in each section of the document. The aim here is for the reader to exercise some discretion: what do you want in such a document. We will look at some of the key elements in this chapter.

8.3.1 Purpose of the Plan

The purpose of this plan is to provide documented intent for the whole validation, such as

- system to be validated
- scope and boundaries of the system
- definition of the life cycle used
- documentation to be produced during each stage of the life cycle
- roles and responsibilities of all involved in the project.

8.3.2 When to Write the Validation Plan?

Ideally, the validation plan should be written as early in the life cycle as possible to define the overall steps that are required and the documents to be produced from each phase of the life cycle. There are different approaches to writing validation plans. You could write the validation plan either as the first or second document in the life cycle. I have no particular preference but I advise you to start writing it after the first or second draft of the user requirements specification to incorporate any implementation or rollout issues in the overall validation strategy. The rationale for this approach is that the validation plan provides documented evidence of intent of the validation.

The document will set out the overall strategy of the validation, defining the life cycle phases and the documented evidence that will be produced in each phase. If you leave writing the validation plan until later in the project, then one or more phases of the life cycle will have passed and you may need to write documents retrospectively and with lower quality as it will be done from memory with few contemporaneous notes.

This is a key document as it defines what you will do in the validation effort and you will be judged against this when inspected. Therefore, read and understand it well – do not write it and forget it because what you write in the validation plan does not always come to pass. There are usually deviations from the validation plan that you will need to document such as reports not written, new ones required that have not been specified or parts of the life cycle omitted or modified. These will need to be noted under the deviation procedure that you have in place in the plan. Although it sounds difficult, once the principles are understood it is relatively simple to do.

8.3.3 Project Plan and Overall Timescales

There is a great temptation to inset an aggressive project plan into the validation plan. This looks good initially and impresses management but when realism kicks in and the timescales are not met it looks foolish. Therefore, simply cross-reference from the validation plan to the current version of an externally maintained project plan or GANTT chart. This can be maintained in a software application such as Microsoft Project and every time the plan is updated the version number of the file is incremented.

It is difficult to give precise timescales for the validation of a CDS as a project varies in both scope and size. However, some broad timescales can be stated from experience as shown in Figure 28:

- Process mapping and optimisation will take between 2 and 3 months to complete.
- System selection should take between 3 and 6 months to complete (this excludes the time to prepare and approve the capital acquisition forms and send the purchase order to the vendor).
- Initial validation (planning, installation, end-user testing, writing procedures and training users) of a new system is dependent on the training of the key

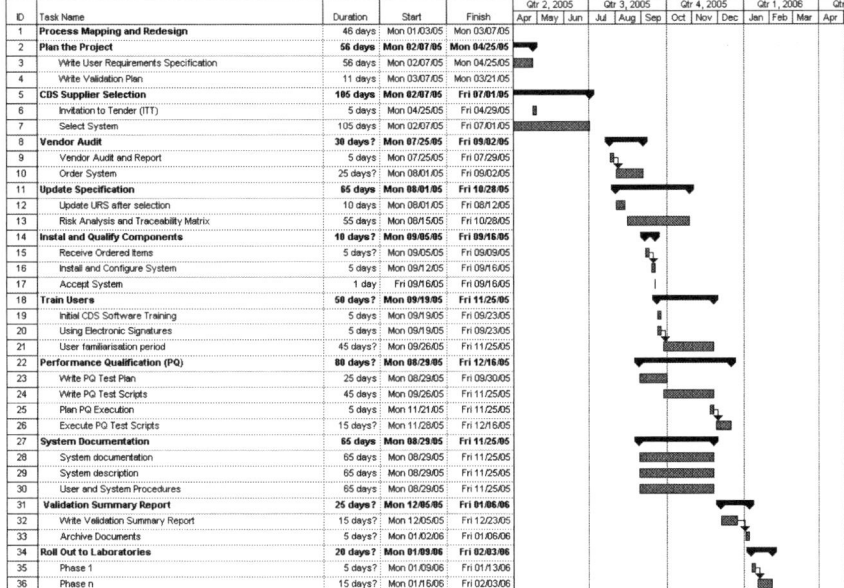

Figure 28   *Outline GANTT diagram for a CDS validation project*

users and their understanding of how the application works; this can take between 6 and 9 months depending on the skill sets of the validation team members.

- If the system is large and there is a phased rollout, then add between 1 and 3
  months to the timescales (*i.e.* an overall total of 7–12 months).
- If the work is an upgrade of an established and validated system, then the
  timescales can be reduced to 2–4 months as the validation documents already
  exist and only have to be modified with the added advantage that the users are
  already trained in the use of the system.

Where CDS validation projects run across departments or between different sites,
an increase in time to the above timescales should be allowed for building and
maintaining consensus of views and approaches.

8.3.4 One Validation Plan for the System Life or One for Each Software Version?

The validation plan, outlined in Section 8.3 also includes maintaining validation
during the operation of the system and can also cover what to do when the system is
retired. This is a comprehensive approach and has the advantage that a single
validation plan for a single system is written. It requires, however, an effective IT
change control system to be effective, as once validated all changes to the system
will be handled *via* change control.

In contrast, an alternative approach is to write a validation plan for a specific
version of the CDS application software. This approach can specify more exactly

what to do for an upgrade of the CDS software; however, it still needs to be linked with the change control system as discussed in Chapter 20.

8.3.5 Roles and Responsibilities

Who is involved in the validation and what do they do is a key section of the validation plan. This can be documented in a number of ways. The easiest way is in a tabular format illustrated in Table 7, where an individual can be identified along with their responsibilities.

There are a number of points to discuss with this illustration:

- The role is not linked to a specifically named individual but a position. With rapid reorganisation, mergers and acquisitions that occur in the pharmaceutical industry, this is probably the easiest course to take.
- The position of the individual named as the system owner allows the role to be traced to an individual.
- The responsibilities of this role are also listed in Table 7. Think about a department head and a small CDS system with 5–10 users then the system owner could carry out these responsibilities. For a larger system with over 100 users, the system owner would be unlikely to perform many of these tasks and would delegate them to a team. For example, with a larger system the system owner would be unlikely to be a trained and experienced user of the CDS and therefore would not execute test scripts.

Ensure that all roles and the corresponding responsibilities are defined for the following list:

- users
- IT department
- quality assurance

Table 7 *Example of the responsibilities for one of the roles involved in a CDS validation*

Role	Responsibilities
System owner: Head of Analytical Chemistry Department	Vendor selection
	Contracts with external suppliers and support services
	Plans and evidence of validation
	Assess the regulatory risk associated with the system
	Documented procedures for users and user manuals
	Release system for operational use
	Operate the change control procedure and the system in a compliant manner
	Execute test scripts

- vendor
- internal validation group
- validation contractors and/or consultants

8.3.6 Validation Team Considerations

One important combination of personnel is the validation team. This is the heart of the whole validation effort and the personnel need to be selected with care. The team members need to have a combination of education, qualifications and experience that allows them to be useful members of the team. They need to include good chromatographers who have an understanding of the application but also need to have the education and experience to enable them to write all the documents throughout the life cycle. This combination can be difficult to find or may place much emphasis on a few individuals. Therefore, the validation team members will be sought after individuals who may be in short supply.

An issue that is not often discussed is in which language should the documents be written? For native English speakers there will be no thought or consideration of this question but those to whom English is a second language this can pose an issue. The ramification is that fluent English writers must be members of the validation team if the decision is made to write the documents in English.

If the project is going to take a long time, laboratory management must change the individuals' position descriptions to include the responsibilities that they will undertake in the CDS validation. This enables the person's performance to be based on their work within the project team as well as in the laboratory.

The project team must be supported by management and allowed to perform the work that they have been required to do, this may also require a project charter. This is especially true when the project is multi-department or multi-site.

8.3.7 Defining Life Cycle Tasks

How detailed does the life cycle information need to be? Each phase of the life cycle needs to be presented with the work to be undertaken with the expected output from each. The problem is always how much detail we need. To help understand one approach to writing the life cycle section in the validation plan, here is an example for the installation qualification (IQ) of the CDS application software:

This activity will consist of:

- *Evaluating the vendor's IQ material to see that it is acceptable*
- *Installing any instrumentation and PC clients according to the manufacturer's guidelines*
- *Installing the software on the server and clients*
- *Executing the IQ: The CDS Vendor's staff will undertake this work and a member of the Laboratory's staff will review the work and confirm that it has been executed correctly with acceptable results*

The aim is not to write a novel but to outline the tasks in sufficient detail that they can be transferred into a project plan or GANTT Chart.

Is this level of detail sufficient?

- The overall work for this phase of the work is defined for this phase of the life cycle. This appears acceptable.
- However, look carefully at the first bullet point; what is missing? The point states that the material will be evaluated but where is the evidence that the work has been carried out? Therefore, what documented evidence is required, and you will need to identify who will be doing the work, the vendor or the laboratory staff?
- Similar questions can be raised with some of the other bullet points. The aim to provide enough detail to identify who will do the work and how will it be documented.

8.4 Defining a Validation Strategy for Larger CDS Systems

One section that is an addition to the IEEE standard is the validation strategy in Section 5 of the validation plan. This is important as many larger CDS systems are not implemented in a single phase but are rolled out over two or more phases or use multiple instances that are to be installed and validated. The purpose of this section in the validation plan is to describe the strategy for the overall validation to ensure overall control.

Larger systems, where there are typically more than 50–100 users on a single instance of the CDS application, will tend to have a phased rollout across the laboratories where it is used. Another situation is where a multinational company standardised on a single CDS application that is installed on multiple instances throughout its facilities. How will these situations be controlled and validated?

In the first situation described in the previous paragraph, an option is to validate an initial instance and as the system is expanded demonstrate that the system continues to perform as intended and defined in the URS. In addition, if a specific laboratory has unique requirements, these can be validated when the system is rolled out in that laboratory. Therefore, a strategy for the overall system validation should be documented in the validation plan. An example could be as follows:

The following validation strategy has been devised:

- *Validation of the initial installation consisting of a core laboratory.*
- *Further rollout of the application of the system to additional laboratories as defined below:*

  Laboratory 2

  Laboratory 3
- *Additional testing of the system will be undertaken if new requirements used by these laboratories before they are rolled-out. Capacity testing of the whole system will be performed before Laboratory 3 goes live.*
- *Expansion will also include the addition of PC processing clients at suitable times. When the same software version and patches are installed, it will be considered an extension of the existing system. The project plan will not record this phase of the work but it will be adequately documented through the change control process for the system.*

Alternatively, if the same CDS application is to be installed as separate instances in several laboratories (global CDS system), the validation strategy should discuss the full validation in one laboratory and reduced validation in the other laboratories as the same instance of software will be used throughout.

The validation strategy should also consider migration of data from previous and existing versions of the CDS systems; this is discussed in more detail in Chapter 23.

CHAPTER 9

System Selection

This section is only applicable for the initial validation of a new CDS system. However, as you will be using a new system for up to 10–15 years, you will need to be confident that the selected system meets your requirements and the CDS will do the job you ask of it.

9.1 What do the Regulators Want?

9.1.1 PIC/S Guidance

PIC/S Guide Section 9.3[31]:

> *The URS, although independent of the supplier should be understood and agreed by both user and supplier.*

This has a link to a footnote at the bottom of the page.

> *Note: This is straightforward for a bespoke system. However, for marketed proprietary systems or configurable packages then it is for prospective users, integrators and suppliers to discuss and review proposed user requirements, versus package functionality. It is essential to determine the 'degree of fit' and then control any necessary configuration work, modification, coding, testing and validation requirements in line with this guidance.*

Section 11.3: Supplier and developer reputations and trading histories for the software product provide some guidance to the level of reliability that may be assigned to the product supplied.

> *The pharmaceutical regulated user therefore should have in place procedures and records that indicated how and on what basis suppliers were selected.*

9.1.2 Regulations Summary

Does a commercial CDS match your existing or proposed ways of working? If not will you make any compromises either in the way that you will work or the way the new CDS will be used? The rationale for selecting a specific commercial CDS system needs to be documented and justified.

9.2 Investment Protection versus Seduction by Technology

The purchase of a new CDS system must be a formal selection process to see if a vendor's CDS application matches the mandatory requirements of your URS. For this reason, the URS is a key document in guiding your selection of a vendor and a system.

Never, ever, select a CDS, or any other computer system, without a prioritised URS, as you will end up being seduced by technology or marketing and might buy the wrong system for the job. You may have failed and wasted money and resources in the process.

9.3 The System Selection Process

We will discuss the options available for selecting a CDS application and vendor; the approaches outlined here will need to be tailored to your organisation and available resources.

9.3.1 Generate a List of Potential Vendors

To select a CDS system and vendor, firstly generate a list of potential suppliers of chromatography data systems. Vendors can be listed through experience, trade magazines or a web search. This raw list needs to be refined into a manageable few that you can send an ITT (Invitation to Tender) or RFP (Request for Proposal) to, as there are many systems available in the market place. Discussions with existing users of these systems can shorten the list. Information on this topic can be obtained through personal contacts and scientific organisations.

The list needs to be reduced to a maximum of 3–5 systems as the brain cannot handle a larger number and still remember the details of each system easily. Therefore, keep the systems to be evaluated to an absolute minimum. Be careful of the way that the system is presented and marketed. Personal comments and experience are worth much more than marketing literature, white papers and hyperbole.

9.3.2 Determine Selection Criteria and Evaluation Tests Now

As your requirements for the CDS are contained in the URS, this document can be used as a basis to design the tests to evaluate the various systems offered by vendors. Can the systems offered meet your requirements especially for the mandatory functions? Using the URS requirements for system selection means that the system selected matches your business needs. Consider the following questions when selecting a system. Do not forget that the tests you use for system selection should also include common problems that you know happen in your laboratory.

- Can the system handle all the tests and samples that your laboratory handles now?
- What happens when samples are switched and you only notice after the analysis?

- Can the system handle data changes with suitable audit trail entries?
- How does a proposed system handle technical 21 CFR 11 compliance issues?
- How can I use the system for electronic ways of working? How are electronic signatures implemented? Remembering that only two signatures are necessary under GMP as outlined in Section 6.1.1.
- Is instrument control an issue or not?
- Do laboratory customers want to review results remotely?
- What other applications will the CDS be interfaced with, *e.g.* LIMS or ERP?
- Resilience of the data system: how will the system prevent you from losing data?
- Is migration of data from an existing CDS technically feasible and practicable?
- Can the system be networked in your environment and to your IT standards?

The basic question you must answer is, will any prospective system perform as *you* want it to? To answer this question, use the URS to prepare a series of selection tests that will allow the system to be selected based on objective criteria and common sense. To do anything else will mean that you are open to seduction by technology. This means that you have no criteria for selection and you are at the mercy of the salesperson and the way they present the system to you. Do not be a fool.

Tests should be devised and performed for each selected CDS. The results of these tests, together with the comments of users on subjective elements of the system, should be used to select a final supplier. These tests should include common problems as it is important to know how a system handles these before you purchase it rather than afterwards.

9.3.3 Prepare the Invitation to Tender/Request for Proposal

The document that is used for a system selection is called either an ITT or a RFP (Table 8). The starting point for the ITT is a prioritised URS. The ITT/RFP is sent

Table 8 *Selected sections in an Invitation to Tender (ITT) or Request for Proposal (RFP)*

Section	Contents
Laboratory description	Brief description of the organisation and the industry sector it is in
	Function of the laboratory and how it helps the organisation meet its aims
User numbers	The number of system users
	Definition of proposed user types with an overview of their roles
Chromatographs	Describe the main chromatographs used in the laboratory: vendor and model numbers, types of detectors to be interfaced
	Is instrument control required or just A/D conversion?
URS sections	Requirements for the CDS are presented as defined in the URS

Table 9 *Example of an invitation to tender document*

Required number	CDS requirement and vendor response	Code
3.3.01	The CDS has the capacity to support 10 concurrent users from an expected user base of 40 users *Vendor response:* *The number of users can be supported by the system provided a 12 concurrent user license is purchased*	Y

Notes to complete the table: A vendor should state if the requirement is available off the shelf, requires configuration or customisation, *etc.* in a column on the right of the table. Therefore, the responses for each requirement should include a code in the right-hand column next to each requirement to clearly indicate whether the product functionality is:

- Y = Available in the current software release and works without any modification
- Con = Available in the current release but requires configuration to function
- Cus = Available in the current release but needs additional custom code (you also need to know who will write this software)
- P = Planned for a release in a later release of software
- N = Not available or not planned for the system

to the selected vendors and a tender should be received within a defined timescale (usually, approximately 4–8 weeks after receipt). If shorter response times are required then alert the vendors that you will send the document to that it is coming so that they can plan ahead.

Note that the last section of the ITT contains the requirements that are copied from the URS. This copy is not exact; as an example requirement 3.3.01 from Section 7.3 is used to show what an ITT section should look like. Table 9 shows the user requirement copied from the URS but without the prioritisation so that a vendor does not put undue bias on the result if the requirement is important. In the document allow space for a vendor to respond. Some guidelines such as maximum number of words for each response should be given to all vendors involved in the tender process to ensure that the response is manageable. Furthermore, to facilitate all vendor responses and to keep them the same, send an electronic version of the ITT to all vendors.

9.3.4 Evaluate the Vendor ITT Responses

The list of mandatory requirements in the URS establishes the base for selecting a particular system and supplier. The selection criteria and tests can be devised from the prioritised mandatory functions. Think ahead, these outline tests can also be used in the PQ test plan and test scripts, allowing reuse of the work in the selection phase.

Tender responses are reviewed against the selection criteria and those systems that meet all, or the majority of mandatory requirements should be selected for actual testing.

9.3.5 Testing Systems Against Your Requirements

Testing two systems allows you a degree of choice in selecting the final system. If more systems are tested, this will increase the choice, but also the time and effort involved. This is particularly important when a small laboratory is considering a CDS. You will not have the time and resources that are available to a medium-sized or large laboratory and may only be able to test one system. However, it is vital that you spend time on this phase to get the decision right, otherwise you will not know whether you have made the correct choice of system until 6–12 months and a lot of effort, later.

When testing a system, select at least one method with the highest degree of complexity that is giving you problems with integration. How will the prospective systems handle this method and will each one give you any advantage over the current system. Construct other tests that will reflect the range of methods of what you currently perform. If instrument control is important, then what makes and models of equipment must the systems control?

If a new process has been designed for electronic ways of working and use of electronic signatures, will the CDS support this process? Alternatively, will you have to change the process or does the CDS offer better ways of working than you had designed?

9.3.6 Consider User Training Now!

Training is a key component of successful implementation of the new system and must be considered as early as possible in system selection. Training requirements need to be discussed with the vendors and modified to any specific circumstances of the laboratory or organisation. Training should not just cover the use of the system but also how to manage and maintain the system.

Training is also a topic for budget cuts. If a laboratory wants to save money, a "train the trainer" approach can be adopted. This may be a false economy with any size of system because only trained users can work effectively.

9.3.7 Visit or Talk with Existing Users

In addition to the hands-on evaluation, talk to contacts you have about the system or ask the vendor for one or two of their users working in the same functional area as you are (API, finished goods, bioanalytical, *etc.*). Consider both avenues. Using personal contacts will give you basic feedback about the system both good and bad, whilst the vendor will not give you the name of a customer that they have a poor relationship with.

You will need to find out what the laboratory thinks of the vendor concerning:

- Installation of the system including IQ and OQ services if necessary
- Training and support

- Consultancy if required for custom calculations or specific implementation requirements
- Support of the system including what happens with software bugs that are detected and fixed

9.3.8 System Selection and Report

The system you select should be based on the practical experience of using it in your laboratory environment or at a vendor laboratory, if time is short or resources for in-house testing are not available.

An alternative approach to the evaluation of equivalent systems can be the use of a methodology such as Kepner–Tragoe. This provides a rational and reasoned decision-making tool if the chromatographers and the rest of the selection team are trained in its use. Finally, a selection report would be the outcome of this phase of the work in which the rationale for buying the system is outlined as well as being part of the supporting evidence for the CDS validation.

However, before you sign on the dotted line you may want to make sure that the software was developed in a quality manner through a vendor audit (Chapter 10) and the contract protects your interests (Chapter 11).

CHAPTER 10

Auditing the CDS Vendor

A vendor is responsible for the majority of the system development life cycle. Therefore, the purpose of the vendor audit is to confirm that the system has been developed according to a quality management system (QMS) and is supported correctly.

10.1 What do the Regulators Want?

10.1.1 Draft FDA Guidance on Part 11 Validation

This withdrawn guidance document notes[23]:

> ... end users should infer the adequacy of software structural integrity ... by ... evaluating the supplier's software development activities to determine its conformance to contemporary standards. The evaluation should preferably be derived from a reliable audit of the software developer, performed by the end user's organization or a trusted and competent third party.

10.1.2 Preamble to 21 CFR 11 Final Rule

> We do not consider commercial marketing alone to be sufficient proof of a program's performance suitability.[17]

10.1.3 PIC/S Guide

Section 11[31]: However, an assessment of the supplier's QMS and recognised certification alone is unlikely to be the final arbiter for critical systems. The certification may very well be inadequate, or inappropriate.

> In such cases, the regulated user may wish to consider additional means of assessing fitness for purpose against predetermined requirements, specifications and anticipated risks. Techniques such as supplier questionnaires, (shared) supplier audits and interaction with user and sector focus groups can be helpful. This may also include the specific conformity assessment of existing, as well as bespoke software and hardware products. GAMP and PDA guideline documents identify a need to audit suppliers for systems carrying a high risk and have detailed guidance on supplier auditing procedures/options.

10.1.4 EU GMP Annex 11

European Union GMP requirements are outlined in Annex 11 clause 11.5[27]:

> *The software is a critical component of a computerised system. The user of such software should take all reasonable steps to ensure that it has been produced in accordance with a system of Quality Assurance.*

10.1.5 Regulatory Requirements Summary

Understanding how the CDS application has been developed and maintained by a supplier is a critical component of the whole life cycle. To get a complete picture of this process a vendor audit is carried out.

10.2 Rationale for a Vendor Audit

The majority of the system development life cycle for a commercial CDS will be undertaken by a third party, the software vendor. This is shown in Figure 12 as all of the operations under the horizontal line. As recommended by the GAMP Guide, Appendix M4[1] a vendor audit should be undertaken to ensure that the software was developed in a quality manner for Category 4 software.

Only the selected vendor should be audited. This is to save time and resources of all involved.

The vendor audit should take place once the product has been selected and the purpose is simply to see if the software has been developed and supported in a quality manner. The evaluation and audit process is very important part of the life cycle as it ensures the design, build and testing stages (which are under the control of the vendor) have been checked to ensure compliance with the regulations. The audit should be planned and cover items such as the design and programming phases, product testing and release, documentation and support. A report of the audit should be produced after the visit.

10.2.1 ISO 9000: Saint or Sinner?

Although many CDS vendors are certified to ISO 9000:2000 of various scopes of certification and will offer you a certificate that the system conforms to their quality processes. The underlying principle, on which ISO 9000 is based, is that organisations that follow documented practices and procedures in a consistent manner are more likely to create products that meet the customer's needs than those organisations that do not follow accepted practices and procedures. However, you must remember that there is no requirement for product quality in any ISO 9000 standard.

Let us explore in more detail the two key elements for ISO 9000.[52] The first is the QMS and the associated documented procedures it covers and the second is the scope of certification open to vendors.

The QMS is universal to all ISO schemes and covers four overall areas:

- Quality policy statement
- Quality manual with overviews of areas such as organisation, roles and responsibilities, training records, quality function, customer complaints, *etc.*
- Written and authorised procedures detailing how the policy and manual will turn into an effective system.
- Internal audits by the quality manager or representatives.

There are three main types of ISO 9000 certification covering the whole or just part of a product or service life cycle:

- ISO 9001: quality assurance in design, development, production, installation and servicing. Conformance to specified requirements is to be assured by the supplier throughout the whole life cycle of a product or service.[53]
- ISO 9002: quality assurance in production, installation and servicing. Conformance to specified requirements during production, installation and servicing.[54] Note that product development is not covered under this scheme.
- ISO 9003: quality assurance in final inspection and test.[55] Conformance by a supplier only at the final inspection and test.

10.2.2 ISO 9001 and ISO 90003

In the case of software, ISO 9001 is not specific enough and this has resulted in the production of ISO 90003 guidelines specifically for software.[56] The reason for this is that with software there is a need to co-ordinate the activities of both the purchaser and the supplier to ensure that the product that is delivered is fit for purpose. In contrast to normal R&D activities, both the supplier and the purchaser have responsibilities for the specification, selection, installation and support of the software product. Of course, if the purchaser ignores their responsibilities then one of the main principles of ISO 90003 simply collapses. This is the quickest way to throw the investment in any software product down the drain.

The uniqueness of software is shown in the fact that the ISO 90003 guideline is double the size of the ISO 9001 document. This is a simple, but empirical method of demonstrating the complexity of the system development life cycle. Therefore, look closely at the ISO certificate from your vendor.

- Which version of ISO is it?
- What is the scope of certification?

Any ISO accreditation scheme is voluntary and not mandatory, unlike GMP. A company can voluntarily enter and leave the scheme. Alternatively, if there are problems, the accreditation body can suspend an organisation's certification or the organisation can voluntarily withdraw from the scheme.

However, it is important to note that ISO 9000 does not guarantee product quality.

A sceptical or alternative view of ISO 9000 is that it produces a poor product with bad processes that are well documented. You should never buy a software application, or any product for that matter, based only upon ISO certification of the company. Hence, the importance of the URS in defining the application you want, followed by time spent to evaluate and select vendors in the market place.

10.2.3 Marketing Literature and Contracts

It is interesting to look at the Dr Jekyll and Mr Hyde approach of vendors to marketing their CDS applications through proposals and protecting themselves through contracts. Compare both the marketing material and the contract and see the differences. This can be very instructive as we can see from the sales response to questions on compliance with GXP Guidelines and their contact terms and conditions.

The example in Table 10 is quoted from a proposal from a vendor with ISO 9001 certification. In the left-hand side column we can see the proposal showing the benefits of the quality approach and the way that regulatory inspection can be facilitated. Compare this with the right-hand side column of the same table that shows the appropriate section from the contract of the same company that states that the company is neither responsible for errors in their own software nor guarantees fitness for purpose.

Table 10 *Comparison of a proposal and contract terms from an ISO 9001 certified company*

The proposal says	The contract says
"Software products are designed to operate with standard computer hardware in the typical laboratory environment. Their design functionally integrates laboratory instruments and results in a unified information architecture that allows for more effective use and control of laboratory data. This enhances the value and utility of the data, especially when measured against regulatory compliance.	*"The company makes no warranties that errors have been completely eliminated from any licensed software. The company makes no other warranties, express or implied, including but not limited to fitness for a particular purpose or merchantability with respect to any licensed software".*
Validating laboratory software is becoming more complex resulting in escalating costs and longer time scales. Because the company has achieved ISO 9001 and TickIT certification, while stressing a high level of commitment to open industry standards and software design, customer validation is made easier. ISO 9001 with TickIT certification can significantly reduce time on customer audits of vendor facilities as well as user functional validation".	

10.3 When do I Audit the CDS Vendor?

Before you purchase is the short answer. In the "honeymoon" period between selecting their system and placing an order the vendor can be very helpful. This may change after the application has been delivered. Therefore, if there are any corrective actions that require holding back money to ensure that actions are completed, it is best to get them onto the table as early as possible in the process. It is difficult to recover from a poor bargaining position, as it will invariably involve more time for discussions and/or legal action.

In parallel to the vendor audit, you should review the vendor's contract terms and conditions, and, as a result, you may also want to negotiate some terms in the contact. Combining the two is good timing and good sense. The contract is discussed in Section 11.2 in more detail.

10.3.1 On-Site or Remote Audit?

The minimum is a remote vendor audit using a checklist that the vendor completes and returns to you. This is usually easy to complete but the writer of the checklist must ensure that the questions are written in a way that they can be understood by the recipient as there may be language and cultural issues here that could impact a remote checklist.

The on-site audit requires a team to visit the vendor's site and review the QMS and how the CDS software has been developed and is being maintained. This takes more effort but the documented evidence can be examined first hand and a better assessment of the overall quality.

10.3.2 Remote Vendor Audit

This is a postal audit that consists of two parts; the first is the generation of the questions by the organisation and the second is the completion of the questions by the CDS vendor. The vendor needs to understand the questions and respond truthfully. However, there is little way of checking the answers you receive. But, for smaller software systems, and some CDS fall into this category, a remote audit is a cost-effective way of getting information on how a vendor carries out their development process as long as you know and understand their limitations.

The remote audit is a cost-effective approach but it cannot ask in-depth questions nor evaluate procedures and evidence. For example, Table 11 shows an example question and answer from a remote audit that due to the nature of remote audits tend to be relatively vague. Note that for both this and the on-site vendor audit, the structure of the document is for the question and answer to be documented together; the questionnaire forms the basis for the report.

10.4 On-Site Vendor Audits

An audit is essentially an independent check of a service or a product; there are three types of audit: first party, second party and third party audits. A second party

Table 11 *Example of a remote vendor audit questionnaire and response*

Question	Response
Q7. Provide an overview of your knowledge of pharmaceutical industry regulations	*50% of the company's sales are to the pharmaceutical industry and therefore we are aware of the GXP regulations as they apply to the laboratory*
What training in pharmaceutical industry regulations do you provide for your staff?	*We provide an overview of the regulations when a new member of staff joins the organisation and to all members when there are significant changes to regulations*

audit is where the user or somebody on their behalf will assess the vendor and their ability to design, produce and maintain a product or service.

The overall purpose in a vendor audit is to assess the quality of software development and what you, as a user, should do to ensure that the system selected and the company that supplies it are suitable. The auditor needs to understand the software development and maintenance process, also a user should be a part of the audit team.

The process is shown in Figure 29; ideally this should run in parallel with the final selection of the vendor:

- Define the scope and boundaries of the system or application.
- Assess the business and regulatory risk of the system.
- Make a decision: if business and regulatory risk is high then a vendor audit is required, if low, then no further action is required or a remote audit will be used.
- Notify the selected vendor of your intention to audit, if they refuse then select another supplier.
- Agree a date and send the audit checklist.
- Conduct the audit and report the results.

10.4.1 The Scope of an On-Site Audit

Having arranged the audit with the vendor, how will you do it? There are three main areas to audit:

- The company
- The quality system
- The product

Normally, a checklist is used as a guide to the audit. Any checklist will have to be customised for each vendor and their associated product, as there may be specific areas to audit. The key to determine what you do is to match what the system or product is going to do *versus* the impact that it will have on your

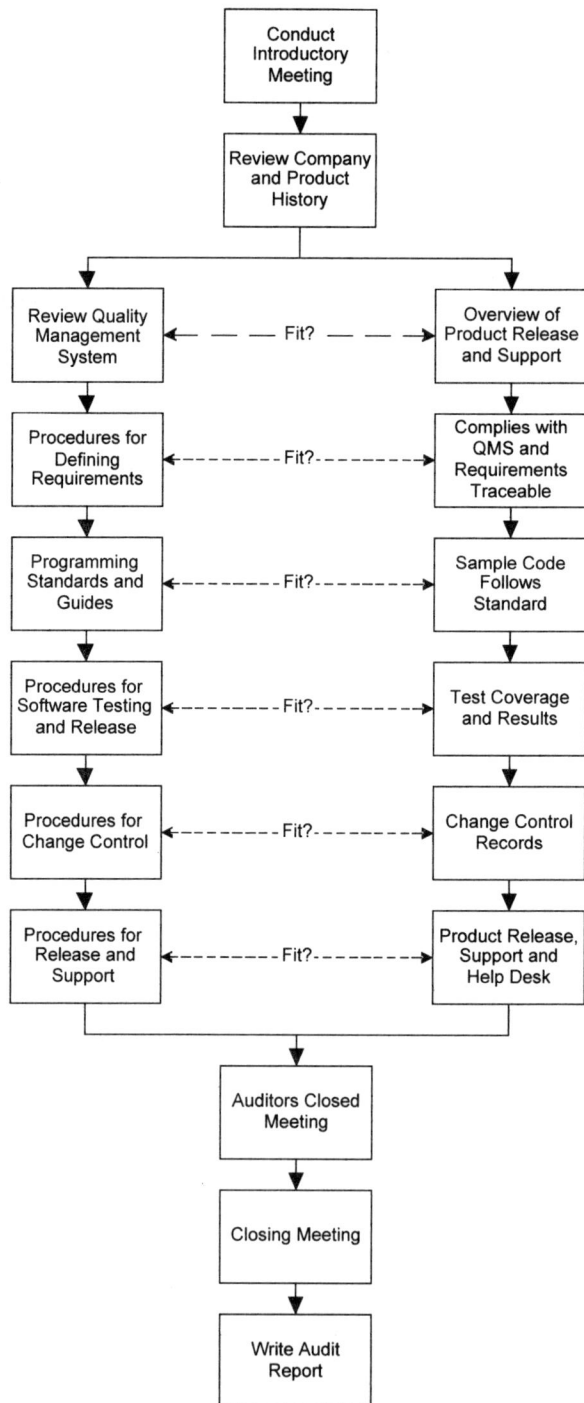

Figure 29 *The overall process for on-site vendor audits*

business. For a critical system, all three areas may be audited, whereas for a lower risk application only the company should be audited.

The coverage for each area is typically:

- *The Company*: This covers general background information such as company history, size, previous experience with the industry in which you are working, written standards and procedures for the life cycle of the product using a defined life cycle model. This can be extended to delivery and installation services, service support after purchase, training of personnel, training services and then look towards escrow services (to ensure you have access to the software if the company goes bankrupt). This type of audit can be used as the basis of a remote audit if some specific product questions are added. This part of the audit can also be a part of the selection process by asking the company to present an overview of itself and approaches to quality.
- *The QMS*: The quality system of the vendor is examined through a series of questions that starts by generally asking about how quality system is reviewed and maintained. If the vendor is certified, the standard and scope of certification should be established with a copy of the certificate. There should be written standards for developing, programming, testing the application coupled with procedures for change control, configuration management and document review. Training of staff is also important and should be documented clearly. Evidence of continuous improvement and evolution of the quality system is very important. A rough rule of thumb is that if the quality system is static – it means it is not working and is just being used for marketing purposes.
- *The Product*: A product audit may look at similar topics to the quality system with the exception that the questions are focussed on a specific product or service. Some overlap with the quality system questions may arise but this is part of the customisation process. Ask here about the programming and structural testing of the product, where individual units and modules of code are integrated together and tested until the final product is ready. If there are different operating systems and hardware platforms supported, ask how much development and testing your version has received compared to other units. You may be surprised to find out how little this is. Manufacture and dispatch of the software, change control, communication of problems and software updates are additional areas to examine.

The overall aim of an audit is to gain the confidence that the company knows what it is doing and that the quality of the product you are purchasing is adequate for the purpose to which it will be put.

10.4.2 The Role of an Audit Checklist

Preparation for an on-site vendor audit is essential as you will usually be limited to the amount of time available and you must concentrate on key areas. In my experience, a checklist is a good way to go, but do not become a slave to it. If there are concerns in critical areas, then follow them and leave some other parts of the checklist incomplete.

Should you give the vendor a copy of your checklist before arriving? There are two schools of thought on this one, yes and no! Personally, I favour being open, as nobody can fabricate a quality system and quality system development documentation in the 2 weeks between sending the checklist and arrival on site. Therefore, I would let a vendor have the checklist as it allows them to prepare and have documentation and people available.

It is important to remember that the aim of the audit is to gain an impression of the quality procedures of the vendor. Note the use of the word "impression". You are getting a snapshot of the process, not an in-depth working knowledge of the vendor's system. To help you draw conclusions as you follow the checklist you will be able to collect evidence (copies of documents, *etc.*), subject to confidentiality of the vendor, of the tasks involved in development of the software product that you are proposing to purchase as you go through the audit. This evidence will help you in preparing the audit report later. Take notes as you go through the meeting. If there are more than one person involved in the audit from the customer organisation, there is always the option to split tasks and cover ground in parallel. Alternatively, if all are involved in the audit process together, it is possible to devise roles for each before the meeting takes place. For instance, the lead auditor conducts the questioning, another can read procedures for correctness, another listen and ask questions as opportunities arise.

However, do not let the vendor run the audit. You are in charge. Some vendors take the opportunity to run the show and can intimidate unwary auditors. Treat such approaches with caution and dig for information and evidence that activities have been carried out according to documented procedures.

Some items for discussion during the audit are:

- Scope of ISO certification: this is available on the certificate held by the vendor (usually framed in a prominent position in their facilities). A copy of the certificate should always be requested. What does the scope of certification? Are all of the activities for the product or service you require included?
- Select 2–4 requirements from your URS and trace them from the concept document, through design and test to the released system. This is very important and essential. It is also instructive to see the quality that is built into the product for the requirements that were selected.
- During the vendor audit, care should be taken to see if there is a procedure whereby management can override the quality system. This can totally negate the quality system, but will be acceptable under ISO 9001 or ISO 90003 as it will be a written procedure. This must be treated with extreme caution.
- Testing to fail: Most tests are designed to pass by vendors. Quality is also determined by testing to fail. If this is not done adequately when you trace a requirement through the life cycle – what is the implication for the whole product?

Once the main part of the audit is over, the team should have time at the company to collect their thoughts, discuss their findings and draw conclusions. This is a private meeting for the auditing team to discuss their findings together before the closing meeting with the vendor.

At the closing meeting the conclusions of the audit team are presented and discussed with the vendor. This is an opportunity to correct any misinterpretations before the report is written and, therefore, is a two-way process.

10.4.3 Writing the Report

The checklist should be the basis for the report; Table 12 shows an example where training records have been examined; the audit can go down to some depth; in this case to specific individuals. Contrast this with the example from the remote vendor audit presented in Table 11.

Decide if the vendor is acceptable and purchase the system, if not either select another vendor or start a process of supplier management.

10.5 Audit Repository Center

An alternative approach to either an on-site or remote audit is offered by the Audit Repository Center (ARC); this has been established by the Parenteral Drug Association (PDA) and uses the voluminous checklist published in Technical Report 32.[57] Qualified auditors will write-up reports and lodge them in the ARC and companies can purchase them and save the expense of an on-site audit. For more information on this subject, the reader is advised to go to www.arc.org.

Table 12 *Example of an on-site vendor audit question and associated observations*

No.	Question	Y	N	N/A	Comments
26	Is the staff training adequately organised and documented?	✓			There is a file containing all staff position descriptions, résumés and training records
					• There was a highlighted organisational chart at the start of each section: you knew which position description was being read
					• Gordon Brown is stated to have a PhD on the organisational chart but does not appear to have one from reading his CV/résumé
					• The two position descriptions that were reviewed were out of date
					• There is not a consistent format for the position description. It was stated that at the start of the year a review was started so that all documents would have the same look and feel

10.6 Using the Vendor Audit to Reduce PQ Testing

A new trend with vendor audits has emerged recently. The emphasis is changing and the purpose is to examine the testing performed by a vendor for a version of a CDS application with the specific aim of reducing the extent and volume of PQ testing by the user.

A detailed URS is essential for this purpose so that the evaluation of the specification of the function and how it was tested is examined. Note that you need to look at the specification as well as the testing as some fundamental errors can still find their way on to the released system. For example, regression analysis algorithms have been found to have the axes reversed for the calculation and plotting of results.[58] Evidence and rationale to justify the reduction of PQ testing must be documented in the vendor audit report.

CHAPTER 11

Contract, Purchase Order and Planning the Installation

Although not often considered in the validation of a CDS, the purchase order is the start of the system configuration. A plan is also required for where to install the hardware (server, data acquisition servers and the chromatographs to be connected to the system) and the contract protects the rights of both the vendor and the customer.

11.1 What do the Regulators Want?

11.1.1 EU GMP Annex 11

Clause 18 states[27]:

> *When outside agencies are used to provide a computer service, there should be a formal agreement including a clear statement of the responsibilities of that outside agency (see EU GMP Chapter 7).*

11.1.2 Regulatory Requirements Summary

There is little in the regulations about the contract. However, it is for business reasons that a contract is important, especially with a new system and/or a new vendor.

11.2 The Contract and Protection of Rights

11.2.1 Rationale for Negotiating the Contract

If the vendor audit, the quote and the CDS software are acceptable, you will be raising a capital expenditure request (or whatever it is called in your organisation) and then generate a purchase order. The quote and the purchase order are a vital link in the validation chain as they provide a connection into the next phase of the validation life cycle: the qualification phase. Before the order is placed, there is a small matter of the contract.

If you never consider a contract just pause to consider the following questions:

- Which organisation writes the contract for the CDS application you are buying?
- If anything goes wrong with the system how are your organisation's rights protected?

The purpose of this chapter is not to attack vendors of CDS software, but to ensure that both parties are protected and enter the agreement on equal terms and as long-term partners.

11.2.2 Overview of the Contract

The contract contents should cover the definition of both parties together with their roles and responsibilities. The scope of the contract should include the system software including any warranties, agreed service contracts with service levels. Remember the aim of the selection process is to get the right tool for the right job. The selection process has selected the system and the vendor, but the job is not complete until the contract has been agreed and signed.

A contract between your organisation and a CDS vendor should define the following:

- ground rules for the acceptance of the system which may include an initial set of deliverables including hardware, software, qualification services and training
- arrangements for on-going maintenance and support of the system including when warranties begin and their duration
- agreements concerning the support of the system through changes and upgrades
- mutual guarantees of rights between the parties
- details of what is expected from both parties at specific stages of the work
- arrangements for payment including any phased payments

Any *pro forma* contract provided by the vendor, if properly drawn up by his lawyers, will be designed to his advantage. The final contract, drawn up by both parties, should divide any risk in agreed proportions. No sentence in a contract is a mere formality, but some have more significance (and are more enforceable) than others. The contract should describe what parties agree to do when things go well, but more specifically, it should also cover what they should do if things go wrong. Table 13 shows the general areas that should be included in a contract negotiation. This approach can be shortened or expanded depending on your exact requirements.

Who should be involved in the contract negotiations? It may come as a surprise to know that staff in your purchasing department are trained to negotiate purchase orders. These people can be very valuable but they need to be involved at the early

Table 13 *Stages in negotiating a contract agreement for a CDS*

Stages in contract negotiation	Work involved
Agree price and payment terms	What are the licensing terms for any software involved?
	Phased payment required for the application, hardware and services if implementation is phased?
	Agree milestones and payment details
Agree delivery and installation schedules	Agree delivery dates
	Is installation to be performed by the vendor or the customer?
	Is qualification to be performed by the vendor or the customer?
	How will approval and review of documents be handled?
	In each situation, agree the prerequisites, any instructions and support to be provided by both parties
Agree maintenance and support arrangements	Annual support charges and service should be agreed and documented
	Level of maintenance and support determined: is this sufficient?
	Are product updates and service packs included?
	Notification procedure for software bugs
	Is there any helpline or Internet support web site?
Identify training requirements	The vendor may provide training in use of its software and availability product as part of the installation or this may be a separate service
	The training requirements and the availability of training programmes should be agreed with the vendor
	On-site training, which should be customised to your operation, may be possible if you have the budget
Obtain vendor assurances	As far as possible, the vendor should be asked to guarantee such factors as performance and instrument control compatibility of the CDS system
	Any assurances given during negotiation should be documented and included in the contract
Prepare contract	All of the above agreements should be consolidated in a formal document to be agreed with the vendor and attached to the purchase order

stages of the work to obtain the best terms possible. The worst time to involve them is at the end when everything has been agreed and you want them to write the contract. Depending on the company culture and procedures, the legal department may also be required to review the contract terms and conditions.

If new technology is being used for the first time there this may incur a high degree of risk and an organisation may put a clause in the contract to withhold a percentage of the system cost until all contractual obligations are completed and the system is working satisfactorily.

11.2.3 Key Clauses of a Contract

The following are some of the key clauses in a contract:

- *The system's future.* Contract terms for upgrades and extensions of a system are often neglected but these are a form of preventative maintenance against obsolescence. The contract must also protect certain rights of both parties. Software is generally protected by a licensing agreement. Nevertheless, if the vendor considers his proprietary software to be confidential, access must be guaranteed to the buyer in the event of a regulatory inspection. The buyer also must have a right to expect that the system will be supported for its anticipated lifespan and the right to system updates at a fair price.

  The personality and the character of the vendor are clearly of major importance. The vendor's personality is demonstrated by his response in day-to-day interactions. In smaller companies, a purchase department may not be available to you and you may be responsible for the full negotiations. If necessary, you may need to use external help, such as a lawyer, to interpret clauses in the contract.

- *Responsibilities.* The major responsibilities and obligations are for payment, warranted performance, maintenance, any upgrading and the continued change in the system as circumstances change. A piece of laboratory equipment carries out a limited set of tasks that can be defined accurately in a contract but as the complexity of the equipment rises so does that of the contract. Some of the above points may only be applicable to data systems, robots and LIMS, where there are upgrades to the computer hardware and application software. With CDS, the user must often buy a continuing service in addition to the product. In this respect, it is quite unlike the purchase of a piece of laboratory equipment, and more like a partnership agreement.

- *Payment.* Terms for payment or withholding payment are a key to any contract. These terms should maximise the incentives to continue the contract relationship. A violation of a contract clause should introduce sufficient discomfort that a good-faith effort will be made to complete any outstanding obligations quickly but the penalties should not be so onerous that one party or the other is tempted to walk away. The costs of all vendor warranties are part of the price of the system. This includes old equipment trade-in or agreements struck to cover such warranties.

- *Performance*. The definition or specification of performance if included can be critical to any contract. What will the system do and what will it not do? Judgements about performance can be relatively straightforward for a piece of HPLC equipment. However, a CDS can be more problematical as it may be carrying out critical tasks. Individual system functions can have many ramifications that magnify the impact of small technical difficulties. Therefore, the real measures of system performance are difficult to judge without a detailed agreement. For example, the performance of a multi-user CDS should include a test of performance when the system is fully implemented and tested under duress.

A good contract should protect both parties against the vagaries of personality and character; however, it cannot be a substitute for either. Honouring a contract requires knowledge, firmness and patience on the part of both parties, even when patience is in short supply. If good faith is lost, the solutions tend to become costly, and unsatisfactory, for both parties.

11.3 Purchase Order: Defining the Initial Configuration

Once the hurdle of the vendor audit is over and the contract is agreed, then you can order the CDS software, data servers or A/D units and any qualification and configuration services from the vendor. The purchase order is important for both validation and business reasons as it defines the initial configuration of the system including instrument, software and documentation.

The purchase order will contain the individual items ordered:

- CDS application software including version number
- number of user licences
- number of the data acquisition servers and/or analogue to digital units
- system documentation
- IQ and OQ services
- consultancy provided by the vendor for implementation of the system, *e.g.* instrument connection, custom calculations, *etc.*
- specialist training services and materials for power users, system administrators, *etc.*
- general training courses and materials for users

All can be confirmed, inspected or tested during the qualification stages of the validation of the system.

11.4 Preparing for System Installation

At the same time as you order the system, you can start to prepare the plan for installing the CDS in the laboratories where the system will be used as well as the IT servers and workstations required for the system.

This is an important document as typically the components will not be thrown into a laboratory and expected to assemble themselves, equally so if the vendor is installing the system will need to know where the components are going. Therefore, you must have a plan of where the components will be sited and which chromatographic instruments will be controlled by the CDS or simply use data acquisition by A/D units provided by the application.

11.4.1 The CDS System Installation Plan

The plan for implementation of the whole system will include both laboratory as well as IT elements and for a large system will need much thought especially if this is an initial installation and not an upgrade.

Some of the sections in a typical installation plan would be:

- purpose and scope of the plan
- roles and responsibilities of all involved parties, for example: users, IT, QA, vendor, contractors or other third parties
- diagrams of the laboratory showing locations of systems and the data servers
- installation of the servers within the data centre or computer room together with any network extensions such as extra sockets, cables or switches, *etc.*
- description of how the installation work will be phased

11.4.2 Laboratory Plan

Usually, there will be an outline plan of what is required for each individual laboratory. For example, there will be an outline plan of the laboratory with the benches, chromatographs, A/D and data servers plus any peripherals. The connection of the data server to the network, *e.g.* network socket number should also be included in the plan.

Figure 30 shows an example of an installation of the CDS components for a single laboratory. However, the network connection of the data server to the network is not shown nor is any additional workstation used for data processing.

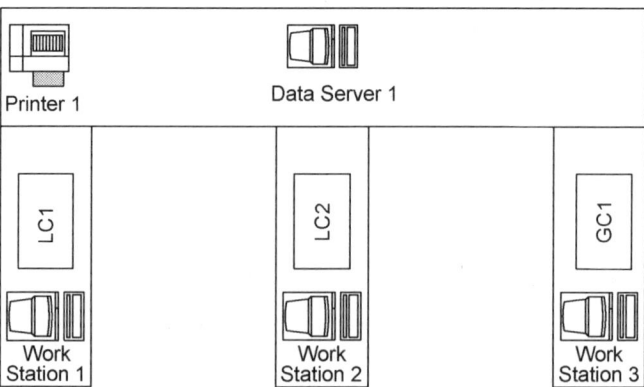

Figure 30 *Diagram of a CDS installation for a laboratory*

This plan can also be linked into the configuration management plan for the system that is discussed in Section 20.2.

11.4.3 IT Plan

The IT portion of the installation plan usually includes information about the servers used for the system (data and application server and terminal servers) plus any network requirements. For example, the network socket numbers and IP addresses that data servers or workstations are connected to must be noted for system installation and support.

To reduce validation costs for CDS systems, terminal servers, *e.g.* Citrix Metaframe, are used increasingly to serve the application to workstations in the laboratories and offices of a facility or between facilities over a WAN. When installed, the client application is installed on a separate server, usually within a data centre, and all processing and interaction occur on this server (or server farm which consists of linked servers). A small applet or viewer installed on each workstation is used to access the CDS application on the terminal server and this saves the installation and IQ and OQ of the client software. With the terminal server approach, the CDS software is installed and qualified once on a server or server farm running the terminal emulation software (*e.g.* Citrix Metaframe). For large systems, the cost of installing the terminal server system is far outweighed by the initial savings with reduced validation and on-going maintenance of the system. In the latter instance, installation and validation of a service pack or software upgrade can be accomplished on the terminal server rather than each individual client.

Figure 31 shows an outline plan for the installation of a database server and the terminal server/farm for a small CDS installation in a computer room or data centre. Missing from this plan are the locations where the servers will be installed

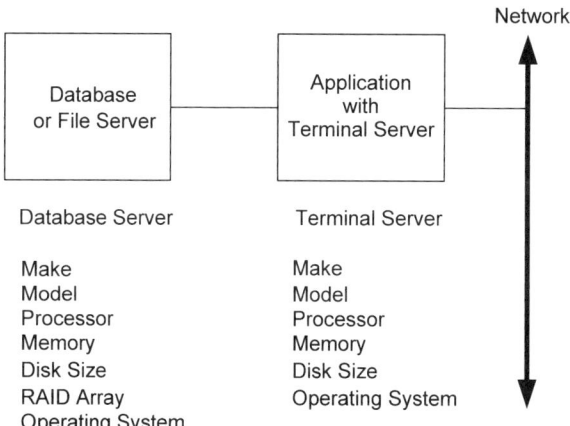

Figure 31 *Outline plan for CDS server installation in a data centre*

(*e.g.* rack and shelf numbers) and also the IP addresses of the two servers and the service packs associated with the operating systems and application. In CDS applications that use a database there are web browser options that can use a separate server that would need to be added to the installation plan.

If terminal servers are used, then the installation of the viewer software or browser onto the workstations used in the offices and laboratories needs to be factored into the installation plan unless these items are standard issue on an organisation's IT desktop.

CHAPTER 12

Risk Assessment and Requirements Traceability

Following from our initial discussion of risk assessment in Chapter 5, the next stage in the process is to carry out a risk assessment of each function to determine if the function is business and/or regulatory risk critical (C) or not (N). This risk assessment methodology uses the tables from the URS that have two additional columns added to them as shown in Table 4.

How requirements are traced to the testing phase and also the validation summary report is also discussed here after the introduction started in Chapter 7.

12.1 What do the Regulators Want?

12.1.1 EU GMP Annex 11

Clause 2[27]:

> *The extent of validation necessary will depend on a number of factors including the use to which the system is to be put, whether the validation is to be prospective or retrospective and whether or not novel elements are incorporated.*

12.1.2 FDA Guidance for Industry: Part 11 Scope and Application

Section on Validation[18]:

> *We recommend that you base your approach on a justified and documented risk assessment and a determination of the potential of the system to affect product quality and safety, and record integrity.*

12.1.3 PIC/S Guidance

Section 4.3[31]:

> *For GXP regulated applications it is essential for the regulated user to define a requirement specification prior to selection and to carry out a properly documented supplier assessment and risk analysis for the various system options.*

12.1.4 FDA General Principles of Software Validation

Section 4.8[29]:

> *Validation coverage should be based on the software's complexity and safety risk – not on firm size or resource constraints. The selection of validation activities, tasks, and work items should be commensurate with the complexity of the software design and the risk associated with the use of the software for the specified intended use. For lower risk devices, only baseline validation activities may be conducted. As the risk increases additional validation activities should be added to cover the additional risk.*

12.1.5 Regulatory Requirements Summary

To justify your extent of validation, your approach needs to be documented with a risk assessment to mitigate and manage the overall risk posed by the CDS within acceptable levels.

From the FDA's comment on the impact of a computerised system on product quality and safety, you must consider "product" as the output from the CDS so that this comment covers both manufacturing and R&D. Too often many people in pharmaceutical R&D see the word product and think that this comment does not apply to them. This is wrong.

12.2 Update the URS Before Starting the Risk Assessment

If a new system has been selected or an existing one is being upgraded then the URS will need to be reviewed and updated to reflect the actual functions that the laboratory intends to use.

12.2.1 Train Key Users

This will be a rate-limiting factor in the overall validation of a new CDS and may also be true if a new version is implemented with significant new features. The key users in the laboratory and those who will administer the system must be trained and be able to understand and operate the new version/system effectively. This will take time and if rushed or done poorly or insufficient depth will adversely impact all subsequent stages of the validation.

12.2.2 Understanding the New System or Version

The URS must reflect intended use of the system by the laboratory, the URS used to select the system will, of necessity, be a generic document. If the CDS selected uses a database then the existing URS may not mention this at all, therefore requirements for the database itself (name, supplier and version number) will need to be added. In addition, the way that an audit trail is implemented in a database (separate table) is different than in a file based CDS. Again this needs to be reflected in the revised URS.

How important is instrument control for a laboratory? Selection of a new system may entail gaining or losing instrument control. Is this fact recognised in your user requirements? Probably not. In addition, you may also decide to take the plunge and implement electronic signatures. This will usually entail significant modifications to an existing or generic URS used to select the system.

12.2.3 Stop Here Until You have a Current URS

Stop.

Do not pass go.

You must ensure that until you have a URS that reflects your intended use of the new or updated CDS you do not go any further with your validation. The URS is the key document around which all further validation work is predicated.

The assumption made is, when you read past this point, that your URS is adequate and has full coverage that reflects:

1. The intended operations of the laboratory and systems that will be interfaced to the CDS.
2. The operations of the version of the CDS to be installed.

12.3 Functional Risk Assessment

As outlined in Chapter 5, there is the GAMP approach to risk assessment[1] using Failure Mode Effect Analysis (FMEA); however, this is considered by the author to be too complex for a configurable application such as a chromatography data system. Therefore, we will discuss Functional Risk Analysis (FRA)[59] which is a simpler and conceptually easier methodology to understand and apply.

12.3.1 Entry Criteria

In this section, we identify, assess and manage the risk posed by individual system functions within the URS. For this reason, the URS must be complete, approved and all requirements are prioritised as either mandatory or desirable.

12.3.2 Risk Analysis of Individual Functions

In the functional risk methodology, each of the prioritised user requirements in the URS tables will be assessed as either critical (C) or non-critical (N) from the perspective of either business or regulatory risk or both.

For a requirement to be assessed as critical the following criteria need to be met:

- It poses a regulatory risk that needs to be managed. The basic question to ask here is; will there be a regulatory citation if nothing is done? For example, requirements covering security and access control, data acquisition, data storage, calculation and transformation of data, use of electronic signatures and integrity of data are areas that would come under the banner of critical regulatory risk.
- A requirement can also be critical for business reasons, *e.g.* correctness of data output and performance of the system. A requirement for the availability of

the system will impact a CDS supporting a continuous chemical production far more than it will be a system in an R&D environment.

All other requirements will be assessed as non-critical.

The approach is shown in Table 14 in the fourth column from the left. Here each requirement has been assessed as either critical or non-critical.

12.3.3 Deciding Whether to Test or Not

The functional risk assessment approach is based on the combination of prioritised user requirements and regulatory and/or business risk assessment. Plotting the two together produces the Boston Grid shown in Figure 32.

Requirements that are both mandatory and critical are the highest risk, medium are those that are mandatory and non-critical or desirable and critical with desirable and non-critical as the lowest risk.

For a CDS, typically most requirements either fall into the high- and low-risk categories. There will be a few requirements in the mandatory and non-critical

Table 14 *Part of a combined risk and analysis and traceability matrix for a CDS*

Req. No.	Data system feature specification	Priority M/D	Risk N/C	Test?
3.3.01	The CDS has the capacity to support 10 concurrent users from an expected user base of 40 users	M	C	TS05
3.3.02	The CDS has the capacity to support concurrently 10 A/D data acquisition channels from an expected 25 total number of channels	M	C	TS05
3.3.03	The CDS has the capacity to support concurrently 10 digital data acquisition channels from an expected 25 total number of channels	D	N	–
3.3.04	The CDS has the capacity to control concurrently 10 instruments from an expected 20 total number of connected instruments	M	C	TS05
3.3.05	The CDS has the capacity to simultaneously support all concurrent users, data acquisition and instrument connects whilst performing all operations such as data reprocessing and reporting without loss of performance (maximum response time is <10 s from sending the request) under peak load conditions	M	C	TS05
3.3.06	The CDS has the capacity to hold 70 GB of live data on the system	D	N	–

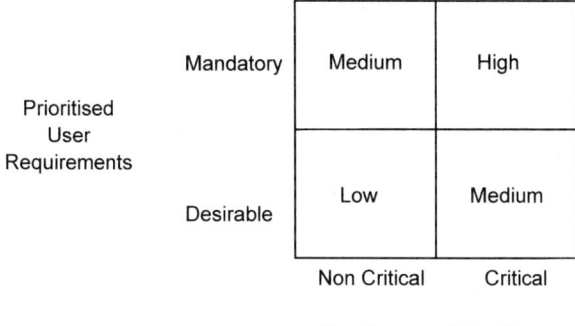

Figure 32 *Plot of prioritised functions* versus *risk assessment*

quadrant of the grid but few, if any, in the desirable but critical quadrant. This is logical. If your requirement were only desirable why would it be critical? If many requirements fall in this last quadrant, it may be an indication that the initial prioritisation was wrong.

The classification of a CDS discussed in Section 5.3.6 is that the basic software is a configurable off-the-shelf application and is GAMP software category 4. Therefore, under this classification, only the software requirements classified as "high" in the grid (mandatory and critical) will be considered for testing in the qualification of the system. No other requirement will be considered for testing.[59]

This approach to risk analysis is far simpler to understand and perform than the FMEA process outlined in the GAMP Guide;[1] it is conceptually easier for end users to understand and apply.

Once we have completed the risk analysis, this now leads us to the concept of the traceability matrix.

12.4 Traceability Matrix

12.4.1 Tracing Requirements to a Specific Test Script

The purpose of a traceability matrix is to show the coverage of testing against requirements. For a small system as a CDS this matrix can be undertaken using the risk assessment by adding an additional column on the right of the table as shown in Table 14.

As outlined in the functional risk assessment, only those functions that are classified as both mandatory and critical are considered for testing in the qualification phase of the validation. Therefore, functions 3.3.03 and 3.3.06 are not considered for testing, as they do not meet the inclusion criteria.

Of the remaining four requirements these all constitute capacity requirements that can be combined together and tested under a single capacity test script, which in this example is called Test Script 05 (TS05). In this way, requirements are prioritised and classified for risk and the most critical one can be traced to the PQ test script.

12.4.2 How Far Should I Trace Requirements?

In the previous section, requirements are traced to specific test script, however it is possible to go further and trace requirements either to a test procedure within a test script or even to specific test step with a test procedure.

Here we are faced with a dilemma that is best described in Figure 16 in the balance between the costs of compliance and non-compliance. It is possible to trace a specific user requirement that is to be tested to a specific test step within a test procedure of a test script. However, it will take time and effort to do this and also to manage the test documentation suite once this precedent has been established. Alternatively, it is simpler and more cost-effective to point to a test procedure within a test script where a requirement is tested and also easier to maintain. In today's risk based environment the second option is a better approach for a commercial CDS application.

12.4.3 Further Tracing of Requirements

As well as linking specific requirements to individual test scripts, a traceability matrix can also be used to link requirements to other deliverables such as standard operating procedures, IQ or OQ documents. In addition, some functions may be specifically excluded from testing as documented in the PQ test plan. For example, even if a vendor were to release its integration algorithm, how good is your math and how will you test it? Other requirements can be verified by linking to the system configuration log such as server requirements or writing procedures.

CHAPTER 13

Installation Qualification and Operational Qualification

Showing that the components of the system have been correctly installed and work as the vendor expects is an important foundation in the overall CDS validation. Typically, third parties including IT and the vendor will carry out much of the work. However, the responsibility of the work planning, supervision and checking remains with the CDS validation team.

13.1 What do the Regulators Want?

13.1.1 EU GMP Annex 11

Clause 13[27]:

> *Before a system using a computer is brought into use, it should be thoroughly tested and confirmed as being capable of achieving the desired results.*

13.1.2 PIC/S Guidance

Clause 13.4[31]:

> *Test scripts should be developed, formally documented and used to demonstrate that the system has been installed, and is operating and performing satisfactorily. These test scripts should be related to the User Requirements Specifications and the Functional specifications for the system. This schedule of testing should be specifically aimed at demonstrating the validation of the system.[22] In software engineering terms satisfactory results obtained from the testing should confirm design validation.*

> Footnote 22 notes. *The supplier/developer should draft test scripts according to the project quality plan to verify performance to the functional specifications. The scripts should stress test the structural integrity, critical algorithms and "boundary value" aspects of the integrated software. The test scripts related to the user requirements specification are the responsibility of the regulated users.*

13.1.3 General Principles of Software Validation

Section 5.2.6 End User Testing[29]:

> *Terminology regarding user site testing can be confusing. Terms such as beta test, site validation, user acceptance test, installation verification, and installation testing have all been used to describe user site testing. For purposes of this guidance, the term "user site testing" encompasses all of these and any other testing that takes place outside of the developer's controlled environment. This testing should take place at a user's site with the actual hardware and software that will be part of the installed system configuration. The testing is accomplished through either actual or simulated use of the software being tested within the context in which it is intended to function.*
>
> *There should be evidence that hardware and software are installed and configured as specified.*

13.1.4 Regulatory Summary

The components of the CDS need to be installed in a controlled way and there should be documented evidence of this work. The system should then be tested to the vendor's specification again with evidence of this activity.

13.2 Terminology: Getting it Right

Before we discuss installation qualification (IQ) and operational qualification (OQ) in the context of CDS validation, we need to look at the definitions of IQ, OQ and performance qualification (PQ) in the contexts of both equipment qualification (EQ) and computerized system validation (CSV). Then the mists will clear and all will be revealed.

13.2.1 Equipment Qualification

EQ is to demonstrate that an item of equipment (*e.g.* gas or liquid chromatograph) is fit for purpose. This implies that all the parameters (*e.g.* wavelength accuracy, linearity of response, *etc.*) utilised by the methods that will run on the instrument are within tested and acceptable limits. Typically, these parameters will use recognised or internationally accepted chemical standards, *e.g.* a holmium perchlorate solution for HPLC detector wavelength. As many methods can be specific to a single laboratory, the instrument parameters to be qualified can vary from organisation to organisation. This is an essential requirement for equipment working in a regulated environment and is the basis for all subsequent analytical method validation work.

As part of the EQ, there are the following stages: design qualification (DQ), IQ, OQ and PQ. These terms are defined in the context of EQ in Table 15.

Chromatographic equipment connected or controlled by a CDS operating in a regulated environment must be qualified. However, a discussion of chromatographic EQ is outside the scope of this book and the reader is referred to the references by Furman *et al.*,[60] Freeman *et al.*[61] and Burgess *et al.*[62] for further information.

Table 15 *Differences in qualification terminology between equipment qualification and computerised system validation*

Term	Equipment qualification	Computerised system validation
Design qualification (DQ)	DQ or User Requirements Specification (URS) that documents the functional requirements of the instrument and any software features including 21 CFR 11 and predicate rule compliance	
Installation qualification (IQ)	Assurance that the intended equipment is received as designed and specified[61]	Documented evidence that all key aspects of hardware and software installation adhere to appropriate codes and the computerised system specification[1]
Operational qualification (OQ)	Confirmation that the equipment functions as specified and operates correctly[61]	Documented evidence that the system or subsystem operates as intended in the computerised system specifications throughout representative or anticipated operating ranges[1]
	Operational release of system	
Performance qualification (PQ)	Confirmation that the equipment consistently continues to perform as required[61]	Documented evidence that the integrated computerised system performs as intended in its normal operating environment[1]
		Operational release of system
On-going validation activities	Change control and configuration management	Change control and configuration management
		Periodic reviews to ensure that the system is still validated and under control

13.2.2 Computerised System Validation

The aim of CSV is to show that the system works and is fit for its intended purpose. This takes a life cycle approach as we have discussed in Chapter 4.6.

However, when we get to the qualification phase of the life cycle, the terms IQ, OQ and PQ are used within the context of CSV also as shown in Table 15. However, their context may differ and this is the major problem with using the qualification terminology. Even the FDA has acknowledged the confusion this causes in the Guidance for Industry on the General Principles of Software Validation[29] and do not mention this approach in this document.

As part of the computerised system includes the chromatograph itself undergoes qualification. Therefore, a user will be undertaking both EQ on the instrument (IQ and OQ) plus CSV on the software (IQ, OQ and PQ). Therefore, your validation plan should make it clear how this problem will be tackled. Will you be qualifying the instrument separately from the software or adopting an integrated approach?

Typically, as most chromatographs cannot operate without the computer and acquire and process data, the integrated approach may be the only option you have open to you.

Note that the monograph by Dyson in this RSC monograph series[3] uses the IQ, OQ and PQ terminology of EQ not CSV. Therefore, please bear this in mind when comparing this monograph and that of Dyson.

13.2.3 Reconciling Equipment Qualification and Computer Validation

Table 15 and Figure 33 illustrate the differences between the way these terms are used in EQ and CSV so that the confusion is minimised. It is important to ensure that the context of the terminology used is established early in a discussion with any other individuals including regulatory inspectors. This is best described in an overall policy document or SOP.

In essence, the DQ and IQ stages outlined in Table 15 are similar between the EQ and CSV terminology and answer the questions: is the system specified and is the system installed correctly? The differences come in the OQ and PQ phases and are due, in part, to the perceived complexity of a computerised system over an analytical instrument. The OQ stage for EQ aims to show that the item is fit for purpose. This is the responsibility of the laboratory and after successful completion the instrument is released for operational use.

With CSV, there are two further stages to go through before the operational release. The OQ is to show that it works as the vendor says it should (anticipated operating ranges) and then the PQ in the user-defined system's actual operating environment. The latter stage is the responsibility of the laboratory. Now the computerised system, once it successfully completes its PQ, is ready for release into a regulatory environment.

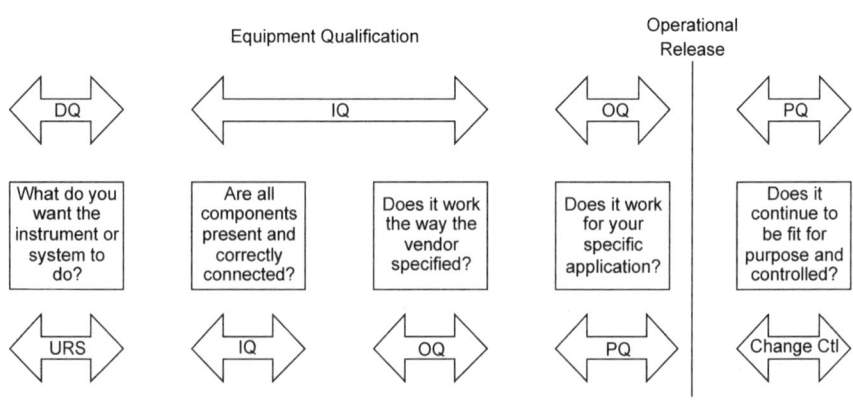

Figure 33 *Differences between equipment qualification and computer system validation terminology (C. Burgess – personal communication)*

What does this mean in practice? Typically, EQ is performed using the CDS to control and acquire data that will support the qualification of individual chromatographs. Also validation of a CDS will incorporate either explicit or implicit testing of the control of chromatographs. Therefore, qualification of the chromatographs and validation of the instruments are closely linked.

13.2.4 Different Aims of Computer Validation IQ and OQ

Having defined the computer validation terms of IQ and OQ what is the purpose of two phases of the system development life cycle? This is shown diagrammatically in Figure 34, at the top is the IQ of a CDS system; there is an IQ of each layer from the bottom up:

- install and qualify the computer hardware
- install, configure the operating system (*e.g.* Internet Protocol address of the computer on the network and operating system functions do you want turned on or off)
- install and qualify the CDS database
- finally, install and qualify the CDS application software

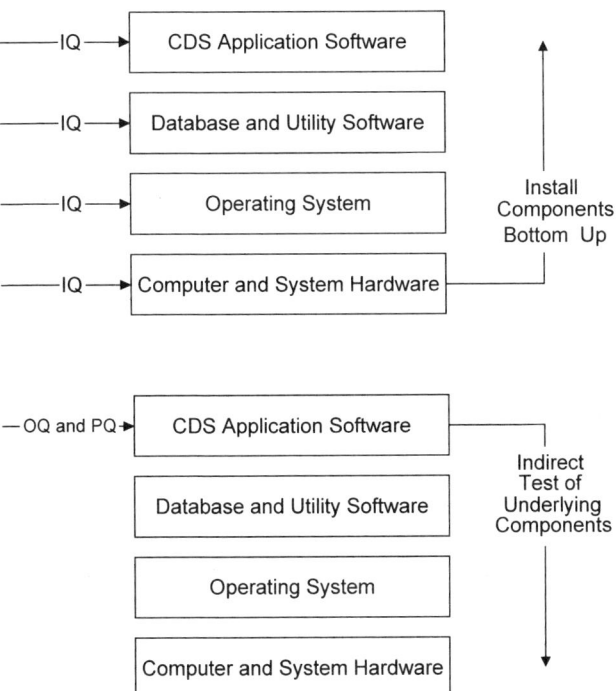

Figure 34 *Differences between installation qualification and operational and performance qualification*

You cannot proceed to the next level up until you have successfully completed the layer below. This is an important concept.

In contrast, OQ (and also the PQ) are only executed on the CDS application layer and base this approach on a successful installation and checkout of the individual components at the IQ stages. Understanding the high level differences between IQ and OQ now enables you to read the remainder of this chapter in context.

13.3 Installation Qualification

Put simply for computerised systems, this is the installation of the components of the system with a check that each works correctly. The best people to undertake this work will be the vendors, as they know their product best. However, there may be several groups working on the IQ depending on the complexity of the configuration of the CDS: vendor, system administrator from the laboratory and IT department.

For stand-alone CDS workstations the following minimum list of activities needs to be completed:

- installation of the chromatograph
- installation of the workstation
- installation of any associated equipment, *e.g.* printer, CD-writer or other backup device
- installation of the CDS software application(s)

The instrument vendor or their approved service agent typically performs these activities or with the workstation there may be a standard build PC provided *via* your company's IT department and then the vendor installs the application software on the top of the operating system and database.

For networked CDS systems the following activities would be required in addition to the items covered under the workstation installation, again depending on the configuration of the system:

- server (for data storage) installation by the IT department, server supplier or manufacturer
- installation of the A/D units or data collection servers onto the LAN
- additional workstations for off-line data processing or review installed by either the IT department or the CDS vendor
- network connection of the workstations to the LAN
- installation of the CDS application software for data processing on the workstations
- connection of the chromatographs to the A/D units or data collection servers

Many chromatographers are not familiar with the detail of the regulations or guidelines that they are operating under. Therefore it is essential to ensure that the documentation to be collected from these IQ activities is planned and collected proactively. Retrospective documentation of any phase of this work is far more costly and time consuming. Therefore, reiterating, plan the work in the validation plan otherwise you will end up with little from this phase of work and a large compliance hole.

13.3.1 Establish the Initial CDS Configuration Baseline Now

The configuration baseline should now be established by doing an inventory of the whole system either during or immediately after the IQ phase of the work. This will result in a description of all the parts making up the CDS including hardware, software and documentation. The configuration records can be kept either on paper or electronically but must be documented. Configuration management is discussed in more detail in Chapter 20.2.

13.4 Operational Qualification

The OQ is carried out after the IQ and is intended to demonstrate that the system works the way the vendor says it will. Most CDS vendors will supply OQ material. These typically will only cover a subset of functions and are not a substitute for the user acceptance tests or PQ tests.

Typically, the OQ is carried out immediately after the IQ and the same person will execute both protocols. However, ensure that before the service engineer starts the work that the individual is trained to do this and you have documented evidence of this such as a training certificate that is current at the time that the work was carried out.

13.4.1 Contents of an Operational Qualification Package

What should be in an OQ? Here this depends on a vendor and the marketing approach to this "value added" package. Here are my views on the subject. The purpose of an OQ is to show that the software and system works the way that a vendor states it should.

To understand the purpose of an OQ more fully, you need to understand how software is produced. As the FDA state,[29] the critical phase of development is the design, writing and testing of the application. Production of the software is simply the production of CD media and verification that the disk has been burnt or copied correctly. Therefore, the main emphasis in software production is the correct design and release of the system. This is where the vendor's certificate (or equivalent) of conformance/validation/compliance with their internal procedures is important. Most of the work is done at the vendor's site and the IQ (have the files been installed in their correct locations) and the OQ (does the software work correctly) are merely confirmation that the software installed operates the same way on your system.

Therefore, the amount of OQ testing can be relatively small, as the vendor has carried out the bulk of the work at their development site. The OQ is merely a confirmation that the "out of the box" software works as expected. Note that no configuration will be carried out, as this is your responsibility.

In most cases, the OQ does not need to be very extensive to demonstrate compliance, especially when the software is to be configured before the PQ is carried out, *e.g.* security, macros, custom calculations, *etc.* The reason is that extensive testing of the baseline package is of little value, as typically the laboratory will change the function by configuration and in some instances customisation (laboratory-specific calculations). Therefore, an extensive OQ is of little overall value especially using a risk-based approach to computer validation.

However, before dismissing any vendor's OQ as a total waste of time and effort, you should, as part of a critical review of the approach, map your requirements to the vendor's package and find out what is being done and can it form a substitute for work you would need to do in the PQ? Some examples may be detailed instrument control functions and where your requirements match what is undertaken in the OQ. Where there is a lot of laboratory customisation of the application, *e.g.* custom calculations or a diode array spectral library using your specific compounds, then the vendor's OQ package is of less or little help here.

13.4.2 Assess the Vendor's Qualification Documentation

Any documentation provided by a vendor must be critically reviewed. Never accept documentation from a vendor without evaluating it and approving it. Why? Let us go back to the regulations; look at the 21 CFR 211 current Good Manufacturing Practice requirements under Laboratory Controls and read § 211.160 subtitled "General Requirements".[12]

> (a) *The establishment of any specifications, standards, sampling plans, test procedures, or other laboratory control mechanisms required by this subpart, including any change in such specifications, standards, sampling plans, test procedures, or other laboratory control mechanisms, shall be drafted by the appropriate organizational unit and* **reviewed and approved by the quality control unit***. The requirements in this subpart shall be followed and shall be documented at the time of performance. Any deviation from the written specifications, standards, sampling plans, test procedures, or other laboratory control mechanisms shall be recorded and justified.*

In essence, you need to have a written plan that is approved by the quality control or quality assurance group within your organisation. How many of you do not do this for vendor-supplied documentation? In fact, how many vendor documents give space for the QC or QA group to sign off that they have reviewed the documentation? However, the regulations go further, much further as we will see from the next section.

> (b) *Laboratory controls shall include* **the establishment of scientifically sound and appropriate specifications, standards, sampling plans, and test procedures** *designed to assure that components, drug product containers, closures, in-process materials, labeling, and drug products conform to appropriate standards of identity, strength, quality and purity.*[12]

A regulation is asking us to be scientifically sound? Now, you see the reason for assessing the vendor IQ and OQ documentation. The regulations require that before execution the protocols have to be approved by the QC/QA unit and also that whatever is written in them needs to be scientifically sound. That is why you must review this documentation and see what you are getting for your money.

Also look at the requirements of the withdrawn draft guidance for industry on 21 CFR 11 validation[23]; in Section 5.4.3, entitled "How Test Results Should Be Expressed", there is the following comment:

Quantifiable test results should be recorded in quantified rather than qualified (e.g. pass/ fail) terms. Quantified results allow for subsequent review and independent evaluation of the test results.

Therefore, this gives you an additional factor for critical review of what you are purchasing. Explicitly stated acceptance criteria must also be available rather than implying if all expected and observed results match, then system passes.

If in doubt the warning letter sent to Spolana,[63] a Czech company, in October 2000, contains the following citation:

Written procedures had not been established for the calibration of analytical instruments and equipment in the Quality Control laboratories used for raw material, finished API and stability testing. Furthermore, calibration data and results provided by an outside contractor were not checked, reviewed and approved by a responsible Q.C. or Q.A. official.

The bottom line is that the laboratory is responsible for the IQ and OQ phases of this work. As much of this work will be carried out by people outside of the laboratory, this must be controlled and documented appropriately.

13.5 Documenting System Configuration and Customisation

When the CDS system validation has successfully completed the IQ and OQ stages of the life cycle, the CDS application software needs to be configured to the operational needs of the laboratory according to the URS. Items to configure include definition of user types and the access control privileges for each, turning on or off of any software switches for 21 CFR 11 compliance and writing report templates. In addition, custom calculations need to be implemented prior to carrying out the PQ.

An alternative way of documenting the items needed for software configuration can be in separate documents but this is a less attractive way as cross-referencing can be difficult to manage.

CHAPTER 14

Performance Qualification (PQ) or End-User Testing

The Performance Qualification phase of the overall validation of the system can be considered as the acceptance testing or end-user testing. This is undertaken by the users and must be based upon the way that the system is used in a particular laboratory. Therefore, a CDS cannot be considered validated simply because another laboratory has validated the same software application. The operations of two laboratories may differ markedly even within the same organisation.

The functions to be tested in the PQ must be based on the requirements defined in the URS and with the numbering of individual requirements can be traced back to the system requirements. The main issue is how can users test their CDS software?

14.1 What do the Regulators Want?

14.1.1 EU GMP Annex 11

Clause 13[27]:

> Before a system using a computer is brought into use, it should be thoroughly tested and confirmed as being capable of achieving the desired results.

14.1.2 FDA General Principles of Software Validation

Section 4.2[29]:

> **Software testing is a necessary activity. However, in most cases software testing by itself is not sufficient to establish confidence that the software is fit for its intended use.**

Note the use of the bold text by the FDA in the original document.

Section 5.2.6 end-user testing:

> User site testing should follow a pre-defined written plan with a formal summary of testing and a record of formal acceptance. Documented evidence of all testing procedures, test input data, and test results should be retained. There should be evidence that hardware and software are installed and configured as specified. Measures should ensure that all system components are exercised during the testing and that the versions of these components are those specified. The testing plan should specify testing throughout the full

range of operating conditions and should specify continuation for a sufficient time to allow the system to encounter a wide spectrum of conditions and events in an effort to detect any latent faults that are not apparent during more normal activities.

Some of the evaluations that have been performed earlier by the software developer at the developer's site should be repeated at the site of actual use. These may include tests for a high volume of data, heavy loads or stresses, security, fault testing (avoidance, detection, tolerance, and recovery), error messages, and implementation of safety requirements. The developer may be able to furnish the user with some of the test data sets to be used for this purpose. In addition to an evaluation of the system's ability to properly perform its intended functions, there should be an evaluation of the ability of the users of the system to understand and correctly interface with it. Operators should be able to perform the intended functions and respond in an appropriate and timely manner to all alarms, warnings, and error messages.

During user site testing, records should be maintained of both proper system performance and any system failures that are encountered. The revision of the system to compensate for faults detected during this user site testing should follow the same procedures and controls as for any other software change.

The developers of the CDS software will not usually be involved in the user site testing. It is important that the user community have persons who understand the importance of careful test planning, the definition of expected test results and the necessity of recording all test outputs.

14.1.3 Draft Part 11 Validation Guidance

There are two pertinent sections in the withdrawn guidance document that cover testing of a system.[23] The first is under General Considerations where Section 5.4 discusses dynamic testing and the second under Commercial Software where Section 6.1.3 that presents functional software testing.

Key testing considerations:

- *Test conditions: Test conditions should include not only "normal" or "expected" values, but also stress conditions (such as a high number of users accessing a network at the same time). Test conditions should extend to boundary values, unexpected data entries, error conditions, reasonableness challenges (e.g., empty fields, and date outliers), branches, data flow, and combinations of inputs.*
- *Live, user-site tests: These tests are performed in the end user's computing environment under actual operating conditions. Testing should cover continuous operations for a sufficient time to allow the system to encounter a wide spectrum of conditions and events in an effort to detect any latent faults that are not apparent during normal activities.*

Functional testing of software:

End users should conduct functional testing of software that covers all functions of the program that the end user will use. Testing considerations discussed above should be applied.

When the end user cannot directly review the program source code or development documentation (e.g., for most commercial off-the-shelf software, and for some contracted software) more extensive functional testing might be warranted than when such documentation is available to the user. More extensive functional testing might also be warranted where general experience with a program is limited, or the software performance is highly significant to data/record integrity and authenticity.

Note, however, we do not believe that functional testing alone is sufficient to establish software adequacy.

How test results should be expressed:

Quantifiable test results should be recorded in quantified rather than qualified (e.g., pass/ fail) terms. Quantified results allow for subsequent review and independent evaluation of the test results.

14.1.4 Regulatory Requirements Summary

From the regulations sections of the guidance and regulations presented above, the following principles for software testing can be derived:

- End-users are responsible for testing the system not the vendor.
- End-users may sub-contract all or part of the PQ to a third party, but typically this is not the vendor as the testing has to be independent and the users typically use a system differently from the vendors design.
- PQ testing must be conducted using the operational system and computing environment. If another test environment is used for a portion of PQ testing this must be documented and any differences between the two environments noted and explained.
- Test the system under all expected uses and operating ranges including some using worst cases especially in a multi-user networked system.
- Conducting tests that are designed to fail is just as important as those designed to pass. This is especially so in situations such as security and access control.

14.2 Principles of Software Testing

The principle of software testing is shown in Figure 35. Testing consists of defining a series of tests from the requirements defined in the approved user requirements

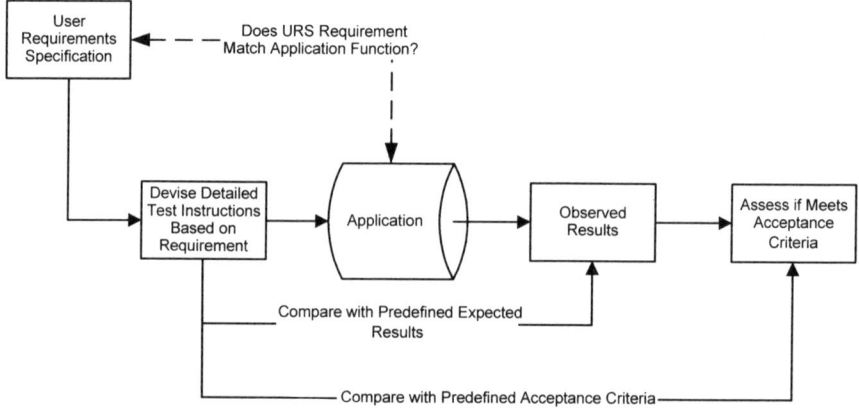

Figure 35 *Principles of test design*

specification. From an individual requirement, a test can be defined for the software to undertake and this test will have one or more expected results and defined acceptance criteria for the test to pass or fail. Remember we will be conducting black box testing (see Section 14.2.3): we usually do not know the detailed algorithms employed by the CDS software only the overall function that the software will perform. At the conclusion of a test, compare the outcome with the acceptance criteria: does it pass or fail?

In devising any software test there should be pre-defined expected results of the test. These will be compared with the observed results made during the execution of the test script. The two should match for the test step to pass. However, there are many test procedures written implying that if expected results match the observed results then the procedure passes. This in my view is wrong, there should be explicitly written acceptance criteria which must be matched for any test procedure to pass.

This is software testing in its simplest form and it is not a difficult concept to grasp, *but* you do not want to test each requirement individually. Elegance in software testing is how many requirements you can test either simultaneously or sequentially in a series of linked test instructions. The quality of testing is not just deriving tests to pass but also tests to fail. Think about common problems and how will the software you are validating cope with them?

Worst case testing does not mean the Martians have landed. Be realistic and look at the system you are trying to validate, *e.g.* computation calculations that place a heavy load on the processor, largest numbers of samples for analysis in a batch and design tests are typical examples of worst case testing.

14.2.1 Testing Approach

One key point is to ensure that the PQ stage progresses quickly, a test script should test as many functions as possible as simply as possible (great coverage and simple design). Software testing has four main features, known as the 4Es[64]:

- *Effective*: demonstrating that the system tested meets both the defined system requirements and also finds errors.
- *Exemplary*: test more than one function simultaneously, where feasible.
- *Economical*: tests are quick to design and quick to perform.
- *Evolvable*: able to change to cope with new versions of the software and changes in the user interface.

14.2.2 Types of Software Testing

Some of the types of testing that could be carried out are:

- *Boundary testing*: the entry of valid data within the known range of a field, *e.g.* a pH value would only have acceptable values within 0–14.
- *Stress test*: entering data outside of designed limits, *e.g.* a pH value of 15 (this is an example of testing to fail; how will the software cope with this data?).

- *Predicted output*: knowing the function of the module to be tested, a known input should have a predicted output.
- *Consistent operation*: important tests of major functions should have repetition built into them to demonstrate that the operation of the system is reproducible.
- *Common problems*: test both the operational and support aspects of the computer system should be part of any validation plan. The predictability of the system under these tests should generate confidence in its operation.

14.2.3 Test Approach: White Box or Black Box Testing?

Therefore as we need to plan our testing, the first issue that we need to face is what are our testing limitations as end-users? Here there are two main approaches to dynamic testing: conventionally known as white box and black box testing and illustrated in Figure 36.

White box testing: This testing requires the full knowledge of what the program unit or module does. This will include the complete specification of the inputs, and outputs and processing algorithms within each module of the software application. The design specification is used to devise tests to prove that the functions described work as designed. In essence, you need to have a programming background to execute white box testing. A normal user will not be able to undertake technical testing either because they do not have the full technical specification of the system and/or they do not possess the technical skills to undertake this type of testing or usually both.

Black box testing: In contrast, in black box testing the tester only knows the overall function of the module or software with the associated input limits. No programming knowledge is required or needed, but training in how to use the application is essential. Therefore, users will undertake black box testing, where known inputs will be entered and the outputs compared with that expected (anticipated results).

In the context of a CDS, one approach to black box testing would be to have known concentrations or amounts of analyte that are labelled as unknown samples

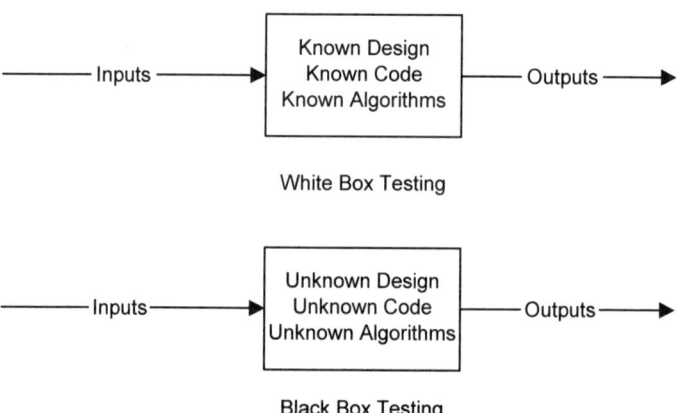

Figure 36 *White box and black box testing*

for the CDS that are measured by chromatographic analysis. Regardless of the test of the CDS you need to have verified data sets against either URS requirements or analytical standards.

So we now know how we will approach the testing of the CDS software.

14.2.4 Manual or Automated Testing?

By the way, if a laboratory is tempted to use an automated test tool to execute their PQ consider Fewster and Graham's words on the subject[64]:

- Automated testing tools take longer to use the first time compared to manual testing
- Expectation will exceed the delivery
- To be economical the test suite must be reused many times (from case study material this is at least 11 times)
- Automated tools are best used for regression testing (to see if operation of the software remains the same after any changes)
- Automated testing is not a substitute for manual testing

Therefore do not use automated testing tools for the PQ, as they will cause more problems than they will solve. However, if a vendor offers an automated tool for the IQ and/or OQ, then this will be useful, as it will establish if the system has been installed correctly and the software functions as the vendor intended it to. However, evaluate any automated software test tool critically to see that it meets your needs and is compliant with GXP as described in Section 13.4.2.

14.2.5 Planning What to Test

The URS should give an explicit understanding of how the software will be used. However, a key requirement in the planning of the PQ testing is training and understanding of how the software actually operates. This will take some time and can be rate limiting in planning the PQ testing. You should ensure that you factor this in when ordering the system so that key personnel can be trained and understand the software as the system is installed on site. Understanding what the software will actually do and how you will use it are crucial to your overall validation effort.

A tempting short cut may be to use only the user manual supplied by the vendor. This is wrong for three reasons.

- The manual does not describe the specific requirements for a laboratory.
- The manual does not define user requirements merely the use of the application.
- The contents of the manual and the software may not match (if features are dropped from the software due to lack of time this may not be reflected in the manual).

Therefore do not use the system manuals as a substitute for the URS for testing the system.

14.2.6 New Data System Features? Update the URS!

As you learn how the CDS software works, there might be features that you did not look at closely before that you will find useful or the business you are supporting needs a different type of analytical support. Therefore, new features and functions may be used; this means that your URS is now out of date and must be updated to reflect the new way that you are using the CDS.

The URS is a living document and needs to reflect the current way you use the system.

14.3 PQ Test Plan

There are a number of ways to document PQ testing; the method that I use which is based on Institute of Electronic and Electrical Engineers (IEEE) software engineering standards and the documentation is derived from IEEE standard 829 entitled "Software Test Documentation".[65] However, there are other ways to approach the problem and you can also use other test plans and protocols. The bottom line is that you must ensure that the testing is documented and covers how you will use the system. IEEE standard 829 lists the main sections of a test plan, shown in Table 16; I use a slight modification of this when I write PQ test plans as not all sections need to be used for a CDS validation.

Once the system to be tested is defined there are three key sections that we will consider in more detail in this chapter:

- *Features to test*: Identifying the test scripts, the features tested in each test script and the requirements tested–traced from the URS. An example of this is shown in Table 17.
- *Features that will not be tested*: This section lists what parts of the system and the software will not be tested and the rationale for this. For example, release notes for the application document the known features or errors of the system but PQ tests should not be designed to confirm the existence of known errors

Table 16 *Outline of a test plan (based on IEEE Standard 829[65])*

1.	Test plan identifier
2.	Introduction
3.	Test system/item
4.	Features to be tested
5.	Features not to be tested
6.	Approach to be adopted
7.	Pass/fail acceptance criteria for all features to be tested
8.	Suspension criteria and resumption requirements
9.	Test deliverables
10.	Testing tasks
11.	Environmental needs
12.	Responsibilities
13.	Staffing and training needs
14.	Schedule (Test order)

Table 17 *Linking requirements to testing in a PQ test plan*

Chromatography data system features to test

Test script identifier	Feature	Requirements tested	
TS01: Logical Security; Access Control and OS Security	Logical security for access to the system *via* the clients will be tested	4.3.1.01 4.6.01	4.6.04
	Password properties will be tested (*e.g.* length, expiration)	4.3.1.02 4.3.1.03 4.3.1.05	4.6.05 4.6.06
	Access by the different levels of user types to specific functions will be tested (*e.g.* administrator, analyst, operator, end-user)	4.3.2.06 4.3.2.07 4.4.11 4.6.02 4.6.11 4.7.01	5.4.01 5.4.02 5.15.07 5.16.01 5.17.03
	Timeout for the CDS system and for the Citrix MetaFrame client will be tested	2.3.07 4.3.1.06 4.3.1.07	4.3.3.14 4.6.10
	The multiple logon system policy will be tested	4.3.1.08	
	Lockout/reinstatement of user accounts will be tested	4.3.1.04	4.6.08

but to test how the system is used daily by the users. If these or other errors were found by the testing, then the test scripts have space to record the fact and what steps were applied to resolve the problem. Furthermore, features such as the operating system and the CDS database will not be explicitly tested, but implicitly tested *via* the application.

- *Assumptions, exclusions and limitations to the testing approach*: As we cannot test everything (as noted by Boehm in Figure 10), we must concentrate on the most critical from both a scientific and a regulatory perspective. When this is done what are the assumptions we have to make, what are the limitations and exclusions to this test approach? This section also provides the contemporaneous notes of testing that can be very useful when inspected as you can refer to it easily to refresh your memory as to why a specific testing approach was taken.

14.3.1 Tracing User Requirements to PQ Testing

You will remember of course, when we wrote the URS that each requirement was uniquely numbered and this means that it can be traced to where in the qualification it is tested. Therefore, you will need a means to trace where an individual or group of requirements are tested in the PQ (or indeed the OQ if the vendor's testing matches your specific user requirements; instrument control may be a good example of this providing you have reviewed the vendor's test approach critically). We will look at this in more detail in the next section covering the PQ plan and test scripts.

14.3.2 Assumptions, Exclusions and Limitations of the Test Approach

As noted by Boehm in Figure 10, complete testing of an application is impossible. Therefore, you will need to limit your testing approach. Section 6.3 of the IEEE test plan is the place to document these. An example of some of the assumptions, exclusions and limitations is shown in the next paragraph:

Test Script 01 – Security and Access Control:

- The functions evaluated in the security and access control test script are not exhaustive and are only representative of the 4 levels of user implemented in the system within the laboratory. The assumption made with this test is that successful execution of the security features of CDS is representative of all functions and all user levels and is based upon the vendor's development of the system and the more extensive testing of these functions.
- A system controller will demonstrate that a user's access rights can be modified in light of being given increased responsibilities.
- A user needs to be registered on both the operating system and the CDS application software. A simple password test will demonstrate that only registered CDS users can access the system. Only the access to the application will be tested and the operating system security will be excluded from testing.
- As the system uses the corporate WAN between the two sites, the test script will check that users in both locations can see appropriate data when this has been set up. The test will take place mid-afternoon, this being the busiest time on the network, and assumes that the test is acceptable for the other times the network will be used by the CDS. The IT Department routinely monitors network traffic and usage. It is assumed that this process will increase the bandwidth before contention becomes a problem.
- When an authorised user is using electronic signatures leaves the workstation unattended for approximately 3 minutes, the system's ability to log them off will be evaluated. The time for the log-off will be timed using a stopwatch but is not expected to be exactly 3 minutes set in the application.

A section such as this will be applicable to all test scripts and for this reason the PQ test plan is not completed until the whole suite of test scripts is nearly completed. This is due to new assumptions, exclusions and limitation to testing being identified as the test script writing proceeds.

14.4 PQ Test Scripts

Under the PQ test plan are the test scripts used to test the CDS application. These are the heart of any PQ testing effort and will take some time and effort to draft and to ensure that they are correct. The concept is that the test script will form the instruction set, the testing log, the archive for the actual testing and all

Table 18 *Outline of a test script (adapted from IEEE Standard 829[65])*

1.	Purpose of the test script
2.	Features to be tested
3.	Test execution instructions
4.	Identification of personnel
5.	Test procedure(s). Each of which consist of the following elements:
	Test steps
	Expected results
	Space to document any observed results during test execution
	Documented evidence
	Acceptance criteria
6.	Test execution log
7.	Test summary log
8.	Testing conclusions and sign-off

experimental data and output will be recorded or logged here. The structure, based on the same IEEE standard as the test plan above, is shown in Table 18.

Each test script (the whole document) can consist of one or more test procedures. Each test procedure typically consists of many test steps or operator instructions on how to perform the test. These test steps and test procedures will test the requirements documented in the URS. We will look at the examples of test procedures in Tables 21–23.

In the same IEEE standard[65] can be found the basis for the test documentation that is the heart of any PQ effort, *i.e.* the test script; in essence this document will contain

- An outline of one or more test procedures that are required to test the CDS functions
- Each test procedure will consist of a number of test steps that define how the test will be carried out
- The expected results for each test step must be defined
- Space to write the observed results and note if the test step passes or fails when compared with the expected results
- A test log to highlight any deviations from the testing
- Sections that collate any documented evidence produced during the testing and include both paper and electronic documented evidence
- Acceptance criteria defined for each test procedure and a statement if the test passes or fails
- A test summary log collating the results of all testing
- A sign-off of the test script stating if the script has passed or failed

14.4.1 Features to Test in any CDS System

From the URS requirements there are three main areas that must be considered for testing any system:

- Scientific and Instrument Control Functions
- 21 CFR 11 Technical Controls and Functions
- Backup and Preservation of Electronic Records

Yes, I know that backup and preservation of electronic records is a 21 CFR 11 issue, but as many people forget to do this when considering a standalone CDS it is given its own specific section as there is more than what meets the eye at first glance.

Going into the CDS in more detail; some of the scientific and instrument control functions that need to be tested are typically:

- Data acquisition
- Calibration methods and calculation of analyte results
- Reporting results
- Custom calculations: if mathematical functions are used for calculations within the system these need to be tested
- Library functions (if used)
- Capacity tests, such as analysing the largest expected number of samples in a single run, especially if the system has an autosampler. This is a specific response to the requirement in the US GMP regulations that the system should have "adequate size".[12]
- Furthermore, tests should be written to demonstrate the handling of common problems e.g. out of range entries are designed e.g. to show the predictability of the system in these areas.

Some of the 21 CFR 11 features to test are:

- System security and access control
- Data file integrity
- Audit trail, especially if the laboratory is working electronically with signatures. This approach is in contrast to the enforcement discretion allowed for audit trails in the 2003 FDA guidance on Part 11[18]
- Ability to discern invalid and altered records

The backup and preservation of electronic records testing are:

- Backup and restore
- Archive and retrieve
- Data migration from older versions of the software or from a different system

Remember one of the key sections of a PQ test plan is the written note of the assumptions, exclusions and limitations to the testing undertaken to provide contemporaneous notes of why particular approaches were taken. This is very useful if an inspection occurs in the future, as there is a reference back to the rationale for the testing. It is also very important as no user can fully test a CDS or any other software application.

14.4.2 Write the Test Scripts

Dependent on the complexity of the system requirements, the overall architecture of the CDS and whether electronic signatures have been implemented, the number of PQ test scripts needed for a chromatography data system typically falls in the range of 15–30 to provide adequate coverage for the important functions documented in the URS.

Typically, the test scripts will cover the following areas of CDS system functionality and are dependent on the defined requirements of your laboratory:

- Data acquisition from the different types of chromatograph interfaced to the system
- Crosstalk of A/D converters[7] or capacity of data servers
- Calibration methods used within the laboratory (are they mathematically correct?)
- Analyte calculation
- System suitability test parameters
- Reporting data
- Sample continuity
- Unavailability of the network (buffering of the A/D or data collection devices)
- Remote processing over the network
- Data acquisition and data processing using a diode array detector (DAD) and/ or dual wavelength detector
- Creation and management of DAD spectral libraries
- Custom calculations implement calculations on data
- Macros used to perform functions automatically
- System capacity tests, *e.g.* analysing the largest expected number of samples in a batch, were incorporated within some test scripts to demonstrate that the system was capable of analysing the actual sample volume that could be expected in the laboratory.
- Interfaces between the CDS and other software applications, *e.g.* LIMS

21 CFR 11 and other regulatory requirements, *e.g.*

- Preservation of electronic records, *e.g.* backup and recovery; archive and retrieve
- Data file integrity
- System security and access control including between departments and remote sites
- Audit trail
- Date and time stamps
- Electronic signatures
- Identifying altered and invalid records

Whilst some of the 21 CFR 11 functions are subject to enforcement discretion under the Part 11 Scope and Application guidance,[18] if you are working electronically as outlined in Chapter 6 then you need to test these functions in the PQ phase of the SDLC.

One issue that needs to be considered is how big should a test script be? There is no simple answer, however too large a test script with over 100 test steps and the collection of many items of documented evidence may be too impractical to use realistically. Equally so a test script with only ten test steps may be too small and carry a high administrative burden to be useful to the overall testing effort. Therefore, a balance needs to be struck in overall size of the test documents.

14.4.3 Outline Test Case Design

The considerations for designing stress and capacity tests for a CDS will be discussed here and will be based on the client–server architecture shown in Figure 7. Note that all requirements for system capacity must be written in the URS and the testing traced to these requirements.

- *Analytical run capacity*: First consider an analytical run and the capacity test considerations that will need to be evaluated. You will know from the URS the maximum number of vials that you will inject in a single run; this will include standards, samples, quality control and blank reagents that you may run as part of your normal procedures. A capacity test should be designed to run the maximum samples anticipated including any replicate injections.
- *Crosstalk*: If two or more channels are multiplexed through a single A/D chip, then a crosstalk test is recommended to see an overloaded signal on one channel impacts another.
- *Data acquisition rate*: Compare the specified data acquisition rate for a data server to the data rate of chromatographs attached to the unit including any diode array detectors. If the total data rate is close to the specification of the unit then test this to ensure that the A/D unit is not compromised during normal operation. If the data rate is far below specification then an alternative path you may decide is that testing is not necessary. However you should document a rationale for this that is scientifically sound. Balancing the regulatory risks is one of the factors in computerised system validation. Do you want to test a function or document the rationale why you did not?
- *Unavailability of the network*: There will be times when the network is unavailable and data will be buffered either in the A/D unit or the data server. You will want to ensure that this function works during the PQ or you will have failed in your due diligence. The worst-case example for the buffering will be defined in your URS and will be the number of injections with the longest run time. The run should be started, then the network is disconnected and the data accumulated in the A/D unit or acquisition unit until the end of the run when the network is reconnected and the buffered data are transferred to the server. There should be no loss of data integrity in any of the buffered and transferred files if this test is to pass. Note also that the date and time stamps must be correct when the network is unavailable.
- *System capacity*: The capacity of the system needs to be tested in a way that reflects on the way the system will be used and there are several approaches to take. If you have a 30 user license then one of the simplest ways of assessing the capacity is to run all systems simultaneously, however, this will only test the data acquisition and transfer to the sever *via* the network.

As the A/D units, buffer acquired data until transferred to the server this test will also implicitly evaluate the transfer with the network traffic at the time of the test. However, one of the main causes of performance degradation will be integration of data and this must also be included as part of any test of system capacity.

* *Logical security and access control*: Whilst logical security appears at first glance to be a very mundane subject, the inclusion of this topic as a test is very important for regulatory reasons as it is explicitly stated in 21 CFR 11. Also, when explored in more depth it provides a good example in the design of a test case.

When designing a test procedure for system security and access control the design could consist of the following elements:

* A single test case where the correct account and password gain access to the system.
* A test sequence where the correct account but minor modifications of the password fail to gain access to the software.
* Three unsuccessful attempts to access the system lead to the user account being locked and a security alert being issued to the system administrator.
* Attempts to change a user password to fewer characters than permitted by the system will result in rejection of the password.

The important considerations in this test design are:

Successful test cases are not just those that are designed to pass but also are designed to fail. Good test case design is a key success factor in the quality of validation efforts. Of the test cases above, the majority are designed to fail to demonstrate the effectiveness of the logical security of the system. The tests rely on good IT practices to ensure that users change or are forced to change their passwords on a regular basis and that these are of reasonable length (minimum 6–8 characters).

Ensure that common problems are included in the PQ test suite. Both the operational and support aspects of the computer system should be part of any validation plan, *e.g.* backup works, incorrect data inputs can be corrected in a compliant way with corresponding audit trail entries, *etc*. The predictability of the system under these tests must generate confidence in the CDS operations (trustworthiness and reliability of electronic records and electronic signatures) and the IT support.

14.5 Defining, Documenting and Testing System Security

Requirements traceability is a key issue in current computer validation best practice. If the requirement is not specified, you have not written the user requirements specification correctly and completely, have you? We will look at system security, as this should be applicable to all CDS systems.

14.5.1 Is the Requirement You are Testing Specified?

The basis for all PQ testing is the system requirements specification and the individual requirements written therein. There is a very simple way to determine if the requirement has been written correctly: can you define a specific test without having to assume anything? If you can, the requirement has been written correctly. If you cannot, the requirement is poorly written and capable of many interpretations.

For example, if the system requirement specification states that "the application must have security functions", this is an example of a poorly written function as explicit tests cannot be derived from it. Instead, more time and effort must be spent defining and documenting the various user types that are necessary and the access privileges allotted to each one.

- *User types*: Typically, there will be a minimum of two: for example, a user and a system administrator/supervisor. Your CDS application will usually have a security module that the system administrator will configure to allow the different user types access to different functions in the application.
- *User privileges*: Any discussion of logical security of an application should consider what each user could do when they use any function. These are the privileges associated with the user of a function within an application. This ranges from the ability to undertake any function to being denied access. These privileges are shown in Table 19 and they are intended to be generic. This continuum may need to be tailored to any specific CDS application in practice.
- There needs to be a separation of the administration activities from the chromatographic ones. For example, if a user is both an analyst and an administrator it would be a best practice to have two user accounts. The analyst account would have no administration functions and the administrator would have no analyst ones. Thus, there is no potential conflict of interest and the analyst and administrator actions traced to the appropriate account.

Table 19 *Continuum of user privileges*

Access privilege	Access rights
Zero-level	No access rights or access denied
Execute only	User can execute functions accessed but nothing else
Read only	User can only read the data accessed but cannot write or append anything
Write only	User can overwrite data
Read–write	User can read or write as required
Append only	User cannot change any data but can add additional information
Administrator	Full access rights to create, read, write, copy and delete data

Table 20 *Defining user privileges for each user type*

Software function	User	Supervisor	System administrator
Account management			
Create/modify/disable user accounts			✓
Create/modify user types			✓
Allocate users to user type			✓
Sample analysis			
Create methods		✓	
Modify methods	✓	✓	
Run samples	✓	✓	
Calculate results	✓	✓	
Electronically sign results	✓	✓	
Run existing report	✓	✓	
Create report		✓	
Electronically approve results		✓	
System backup			
Backup system to tape			✓
Recover from tape to system			✓

The functions available to each user type need to be documented as either part of the system requirements specification or reference made to an SOP where this information is located. One way of accomplishing this is to define the requirements in the form of a matrix of function *versus* user type as shown in Table 20.

Once your system is operational the system administrator should review access rights of all user types on a regular basis especially when they are trained or are promoted which could result in a change of user privilege.

14.5.2 Designing the Tests

Now that we have the requirements for access control correctly specified, we can now start to design the tests that we can execute to demonstrate that the system has been correctly specified and configured.

Look at the definition of access rights for each user type in Table 20. You can define a series of tests directly from this table:

- A system administrator can perform all account management and system backup tasks (testing to pass).
- A user cannot perform account management or system backup functions (testing to fail).

- A supervisor can perform all sample analysis tasks (testing to pass).
- A user can perform sample analysis including modification of methods and running existing report (testing to pass) but cannot create reports or methods (testing to fail).

If the requirements are correctly specified, it is relatively easy to define tests to demonstrate that they function correctly.

14.5.3 Risk Analysis: Extent of Testing?

The question that some of you will ask is "how much is enough"? Others will ask the corollary, "what is the minimum I can get away with"? This is where you need to look at the risk involved with the system and the time taken to test the function. The black and white approaches can be summarised as:

- The simplest way is to test everything: all user types and all functions. No regulatory comeback here, but no real work can be carried out either.
- Companies that come from the Clint Eastwood School of Computer Validation will assume that the vendor has tested everything and do nothing.

Working in the laboratory, you have to consider the GMP regulations: 21 CFR 211.160(b) requires that any work has to be *scientifically sound*. Therefore, consider the documented arguments that you can produce that can reduce the work of testing the access control functions:

- The vendor has tested the basic software system (this is a better argument if backed up with a vendor audit report).
- You have configured the software within the boundaries defined and tested by the vendor of the software (configuration has been documented as described in the previous section).
- Therefore, only test your configuration of the system.
- Do you test all or just representative functions of the system? Even for high-risk systems, I would suggest that you only test representative functions as you are working within the overall framework that is provided by the vendor.

14.5.4 Refining the Test Design

Taking the view that we will only test a representative selection of our configured user functions we can now develop two test cases, *e.g.*

- A system administrator can create a new user but cannot create a method, analyse a sample and create a new report.
- A user cannot create a new user, method or a report but they can analyse a sample.

Now we have designed the test cases, we now have to consider how to document them.

14.6 PQ Test Documentation

Remember that the terminology used here is derived from IEEE software engineering standards; in your organisation similar documents may be called something else.

14.6.1 Key Test Script Sections

From the discussion in section there is a Section 14.4 on PQ testing; there were a number of criteria for testing such as:

- Test procedure steps
- Expected results
- Observed results
- Acceptance criteria
- Documented evidence

A test procedure should consist of three main elements:

- Test steps and expected results (defined and approved before testing starts); observed results, note log, pass–fail statement and who performed the testing (written contemporaneously as the testing is conducted).
- Documented evidence (not specifically required by the IEEE standards but essential for collating information used to support the testing for QA and regulatory inspectors review).
- Acceptance criteria; these must be explicitly stated and *not* implied as they are in many qualification documents.

14.6.2 Documenting Test Execution Instructions and Expected Results

I am often asked questions about how much detail needs to be put into the test execution instructions? This depends on your company approach to risk and the amount of effort they wish to expand. From my perspective, the test steps should be written for a trained user and not a novice to execute. This saves writing the embarrassing series of instructions starting with, for example, sit in front of the workstation, press the on button and wait for the operating system to boot ...

Therefore, write the instructions so that a trained user can execute them. If a user can come to one and only one result, then the test steps have been written correctly. If not they need more detail. The detail depends if this is a new system being validated for the first time or an existing system being upgraded and (re)validated. Often these two will be different as the user's maturity and

Table 21 *Test script execution instructions written for a new user of an application*

Test steps and expected results

Test steps	Expected result
1. The system administrator defines user "Security" as user type "Guest"	User "Security" defined as "Guest"
2. Log onto data system as user "Security"	Access to the system
3. Enter "Configure System" and select "Users"	Function window opens and lists users
4. Select user "Security" and select "Properties"	The properties window for user "Security" opens
5. Check in the General tab the User Type field	User type "Guest" displayed
6. Cancel to leave the properties window	Leave properties window
7. Attempt to access "Select 'New Method'" from the "Method" menu	Function not available
8. Exit "Configuration Manager"	"Configuration Manager" window is closed
9. Enter "Browse Methods" and select a project Name: _____	The window for the specified method opens
10. Verify that the selected method contains results	Method contains results
11. Select a result and select "Preview" in the drop down menu	A window opens, named "Open Report Method"

---

experience differs. Tables 21 and 22 itemise the differences between these two approaches.

Looking at the test execution instructions in Tables 21 and 22, one is written in a very terse style for an experienced user and the other in more detail. Which style is appropriate for your laboratory? This depends, the detailed style is good if you have staff turnover and want to retain consistency of execution. The disadvantage is that if the user interface changes then you could face extensive rewriting of the test script procedures to reflect this.

The terse style has the advantage that it is quick to write but also unfortunately easy to assume implicit tasks that are obvious to the writer but not to the person executing the procedure. The issue is can the implicit steps be remembered the next time the test script must be executed again? This approach is best suited to a multi-user environment where there is sufficient experience to ensure that the scripts are understandable by the user base as a whole rather than just an individual.

Table 22 *Test script execution instructions written for an experienced user of an application*

Test steps and expected results

Test steps	Expected result
1. The system administrator defines user "Security" as user type "Guest"	User "Security" defined as "Guest"
2. User "Security" logs onto the system and looks at his access privileges under "Configure System"	User privileges of "Guest" displayed
3. Attempt to access "Select 'New Method'" from the "Method" menu	Function not available
4. Enter "Browse Methods" and select a project Name:	The window for the specified method opens
5. Access one of the results in the selected method and report result	A window opens, named "Open Report Method"

Regardless of the approach used, ensure that the test scripts are signed and approved *before* they are executed.

14.6.3 Writing Observed Results

When the test steps are executed, we need to know:

- Who executed the test?
- When was the test executed?
- What were the observed results?
- Did the test step pass or fail?

Therefore, these need to be incorporated into the test procedures outlined above.

14.6.4 Unexpected Results

It is rare that a PQ test suite is executed without some issues arising. These can come from a number of causes:

- Misunderstood test instructions
- Wrongly written test instructions
- Insufficiently formulated test pre-requisites or test set-up
- Expected results poorly described

Table 23 *Test script execution instructions with an associated test log*

Test steps and expected results		Test log			
Test steps	*Expected result*	*Observed result*	*Note log*	*Pass/ fail*	*Initials/ date*
1. The system administrator defines user "Security" as user type "Guest"	User "Security" defined as "Guest"				
2. User "Security" logs onto the system and looks at his access privileges under "Configure System"	User privileges of "Guest" displayed				
3. Attempt to access "Select 'New Method'" from the "Method" menu	Function not available				
4. Enter "Browse Methods" and select a project Name:	The window for the specified method opens				
5. Access one of the results in the selected method and report result	A window opens, named "Open Report Method"				

- Wrongly set acceptance criteria
- Software errors found

Regardless of the cause, the test script needs to have a way to document these issues and resolve them. This is where the test execution log is useful.

14.6.5 Suggested Documentation

Table 23 shows an outline that could be used to integrate the test steps, the expected results and the four requirements in Chapter 14.6.3. The level of detail that you should go to is a balance between time and effort. Regardless of the depth of documenting the testing, I strongly suggest that you write the documentation not only for user execution but also for internal audit and external inspection. This effort will repay itself repeatedly.

14.6.6 Documenting Observed Results

This is an area where there is a great variation in approaches by companies depending on their approach to risk, company culture and QA culture. The options here are:

- Write the observed results down verbatim (*e.g.* observed results for test step 3 in Table 23 could be recorded as "function not available") which is relatively laborious but is easy to follow and review after the test step.
- Record observed results as "as expected". This is a simpler way to record observed results but this depends on the quality of the expected results.
- Change the observed results column heading to "Does Observed Match Expected?" Now you can record the observed results as either yes or no.

These are the main options and you make your choice and have to live with it.

14.6.7 Collating Documented Evidence

During the testing of the software, there will be output from the system either in the form of paper printout or electronic records, or for some test steps, both. These need to be retained and collated together as evidence that the test script was executed. To aid audit and inspection, there should be a section in the test script to collate the documented evidence as attachments to the test script.

14.6.8 Has the Test Passed or Failed?

All tests that are performed in the PQ must have explicitly written acceptance criteria. This is a common weak point with most test scripts or protocols. If the test acceptance criteria are not written down before execution, how does a tester, the reviewer or even an inspector know if the tests have passed or failed? The unwritten rule is that if the observed results match the expected results then the test has passed.

Document and approve the acceptance criteria for each test and compare with the actual ones in a specific section of the test script is the take home message here. For example, an acceptance criterion for passwords and user identities could be: only the correct combination of user identity and password will permit an authorised user to access the CDS application. All others will be rejected.

Your test scripts will typically consist of one or more test procedures each with their own acceptance criteria. At the end of the document summarise the results of each of these and state if the test script passes or fails and then sign it.

14.7 Some Considerations for Testing Electronic Signatures

In Chapter 6, the use of electronic signatures was discussed from a process perspective. The documented user requirements are tested in the PQ phase of the life cycle. Listed below are some of the test cases that could be considered for inclusion in the PQ test scripts covering electronic signatures. The pre-requisite for the testing is that appropriate users have been defined in the CDS together with electronic signature signing authorities allocated to them. In addition, samples have been analysed and there is a report available to sign.

Test cases are as follows (note that most are designed to fail rather than pass):

- The appropriate tester and reviewer both sign in the correct order should be supported by the CDS together with documented evidence of the correctly signed report
- Signature manifestations of both the tester and the reviewer meet 21 CFR 11 requirements
- Reviewer attempting to sign as the tester should fail
- Tester attempting to sign as the reviewer should also fail
- The same person signing both as a tester and a reviewer should be rejected
- Attempt by an unauthorised user to sign the report should be rejected by the system
- A reviewer attempts to electronically sign the report before tester will be rejected

CHAPTER 15

User Training and System Documentation

Successful installation and implementation of a CDS requires effectively trained users using good documented procedures. Vendor documentation also provides reference material that will be outside the scope of standard operating procedures (SOPs). Together these contribute to the controlled function elements of computerised system validation described in Figure 9.

15.1 What do the Regulators Require?

15.1.1 EU GMP Annex 11

Clause 1[27] states:

> It is essential that there is the closest co-operation between key personnel and those involved with computer systems. Persons in responsible positions should have the appropriate training for the management and use of systems within their field of responsibility, which utilises computers. This should include ensuring that appropriate expertise is available and used to provide advice on aspects of design, validation, installation and operation of computerised system.

15.1.2 FDA 21 CFR 211 GMP

211.25 Personnel Qualifications (a)[12]:

> Each person engaged in the manufacturing, processing, packing, or holding of a drug product shall have education, training, and experience, or any combination thereof, to enable that person to perform the assigned functions. Training shall be in the particular operations that the employee performs and in current good manufacturing practice (including the current good manufacturing practice regulations in this chapter and written procedures required by these regulations) as they relate to the employee's functions.

> (c) There shall be an adequate number of qualified personnel to perform and supervise the manufacture, processing, packing, or holding of each drug product.

15.1.3 FDA 21 CFR 58 GLP

Section 58.29 Personnel[14]:

> *(a) Each individual engaged in the conduct of or responsible for the supervision of a non-clinical laboratory study shall have education, training, and experience, or a combination thereof, to enable that individual to perform the assigned functions.*
>
> *(b) Each testing facility shall maintain a current summary of training and experience and job description for each individual engaged in or supervising the conduct of a non-clinical laboratory study.*

15.1.4 Regulatory Requirements Summary

Appropriately trained users of the CDS are essential in any regulated laboratory; as noted in Annex 11 this includes the IT staff. It also makes good business sense as well.

15.2 Personnel and Training Records

All involved with the selection, installation, operation and use of a CDS should have training records to demonstrate that they are suitably qualified to carry out their functions and maintain them. It is especially important to have training records and curricula vitae/résumés of installers and operators of a system as this is a particularly weak compliance area.

Major suppliers of CDS will usually provide certificates of training for their staff who will be involved with the installation of the system and software as evidence of their training. However, a problem area for many CDS is that the IT personnel maintaining the system do not have training records or curricula vitae.

The types of personnel involved that could be involved in a CDS validation are:

- *Vendor staff.* Personnel who were responsible for the installation and qualification of the data system software should leave copies of their training certificates listing the products they were trained to work on. These should be checked to confirm that the certificate(s) were current at the time of the work was carried out and covered the relevant products worked on. These copies should be included in the validation package.
- *System managers.* Training given to these individuals in the use of the system and its administration, typically provided by the vendor, must be documented in the individual's training record.
- *Users.* Chromatographers trained to use the system either by the vendor or an internal trainer should have the fact documented in their training records. If the latter approach is to be used then records of the trainer's ability to train users need to be available.
- *Consultants.* Any consultants involved in aiding a validation effort must provide a curriculum vitae (résumé) and a written summary of skills to include in the validation package for the system (this is an explicit FDA GMP requirement outlined in 21 CFR §211.34).

- *IT staff.* Training records and job descriptions outlining the combination of education, training and skills that each member has that are relevant to supporting the CDS, for example, backup and recovery or database administration.

Training records for CDS users are usually updated at the launch of a system but can lapse as the system becomes mature. To demonstrate operational control, training records need to be reviewed and updated regularly especially after software changes to the system. This needs to be part of the change control impact assessment discussed in Chapter 20. Error fixes do not usually require additional training. However, major enhancements or upgrades should trigger the consideration of additional training. The prudent laboratory would document the decision and the reasons not to offer additional training in this event.

To get the best out of the investment in a CDS, periodic retraining, refresher training or even advanced training courses could be very useful for large or complex installations. This additional training should be documented.

15.3 System Documentation

The documentation supplied with the CDS application or system (both hardware and software), user notes and user SOPs will not be discussed in depth as it is too specific and also depends upon the management approach in an individual laboratory. However, the importance of this system specific documentation for validation must not be underestimated.

Keeping this documentation current should be considered a vital part of ensuring the operational validation of any computerised system. The users should know where to find the current copies of documentation to enable them to do their job. The old versions of user SOPs, system and user documentation must be archived.

15.3.1 CDS Vendor Documentation

Vendor documentation provided with the CDS will include all or some of the following:

- User manual (specific to a version of the CDS application)
- Quick reference guide
- System administrators guide
- Release notes for a specific version and any service packs (useful to see what has changed since the last version with new features and which software features or bugs have been fixed)
- Chromatographic integration: theory and practice using the application
- Technical notes
- White papers about specific issues (*e.g.* use of electronic signatures or 21 CFR 11 compliance)
- On-line help for the application

The trend today is to provide the manuals mainly as electronic versions on CD or *via* download at a vendor's web site. One copy of the manuals for each version of the software used needs to be archived with the validation documentation.

15.3.2 Laboratory Standard Operating Procedures

SOPs are required for the operation of both the CDS applications software and the system itself. As explained above, we will not consider user SOPs in detail. SOPs are the main medium for formalising procedures by describing the exact procedures to be followed to achieve a defined outcome. SOPs must be used to document routine operations to ensure that they are undertaken consistently. A written procedure means that new employees are trained faster.

Laboratory staff are used to working with SOPs. However, if a large system is supported by a central computer group they may not be used to working with SOPs and even less ready to document their work. However, to provide a service to a regulated laboratory, an IT department must provide a suitably documented procedure.

One specific SOP for all GMP laboratories using a CDS, especially following the Barr ruling, is on manual reintegration. This is a key procedure. If a laboratory is assaying material using a registered chromatographic method and an analyst needs to manually reintegrate peaks for active compound then the method could be said to be out of control. In contrast, measurement of impurities at or near the limits of detection will usually require manual integration. Therefore, this SOP is very important in describing the situations when manual integration can be used.

Some of the SOPs required for a CDS from the perspective of the chromatographers including the laboratory system administrator are:

- *Preventative maintenance and calibration.* Describing the procedures for any preventative maintenance or calibration of the hardware components.
- *Error logging and correction.* The instructions for finding, recording and resolving errors in the system. This can be a complex SOP covering many different aspects of the system and may refer to sections of the technical manuals provided with the system. If a software bug is found it may link with the change control SOP for the system.
- *User account management.* The procedures for adding, modifying and disabling users; associated with this will be the definition of user types and the allocation of an individual user to a specific user type (this SOP may be the responsibility of the IT department in some larger organisations).
- *System start-up and shutdown.* This is a special SOP that should contain all the specific instructions for starting up and shutting down the system. This SOP may be required in an emergency and therefore should be written well and be easily available for use.
- *Installation and update of software.* Procedures to be undertaken before, during and after installing software. This should start with the complete backup of all disks and then installation of the software and any testing and validation that may be required.

15.3.3 Checking the SOPs during the PQ

Before the PQ is completed, the SOPs for the system (user, administrator and where necessary IT) should be walked through the respective processes to confirm that they reflect actual practices. If not then they need to be updated to match working practices. Checking the SOPs during the PQ means that users will be trained to use the system and will understand what to look for. Although it is an obvious statement, the individual checking that the SOP is correct should not be the author of the procedure.

15.4 Administrative and Procedural Controls Required for 21 CFR 11 Compliance

There are a number of administrative and procedural controls required under 21 CFR 11 to be considered. This list is not intended to be exhaustive but is focussed on the procedures needed for a CDS operating a regulated laboratory and using electronic signatures.

Although not a procedure, it is important to check that your organisation has sent a letter to the FDA certifying that when electronic signatures are used they are the legally binding equivalent of handwritten signatures as required by §11.100 (c),[17] especially if you are going to use electronic signatures in the operation of your CDS.

- *Using electronic signatures with non-repudiation.* An SOP is essential to describe the significance and responsibilities associated with using electronic signatures. This must include non-repudiation (a legal term that means an individual denying that the signature is theirs) in terms of individual responsibility and also the consequences of falsification both for the pharmaceutical organisation and for the individual. Associated with this SOP needs to be a short but effective training course in its application. Both the SOP and the training should include an explanation of the difference between signing an electronic record and identification of actions.
- *Password.* The requirements for a secure password, for example, a minimum length of six alphanumeric characters that are changed every 30–90 days. The procedure needs to state explicitly that users do not divulge or write down their passwords.
- *Change control and configuration management.* Change control and configuration management procedures must be in place that control and document all changes to the system. An impact evaluation and authorisation is carried out before all changes (with the exception of emergency changes). Configuration items and the configuration baseline are defined as described in Chapter 20. There needs to be a QA review of these records, typically annually. Service engineers must be prevented from installing software upgrades as part of routine maintenance without proper authorisation.
- *Changing and monitoring date and time stamps.* This procedure is a critical one for the integrity and trustworthiness of electronic records. Who is allowed to change date and time stamps on the system? How will any checks and

changes be recorded? Any changes including summer/winter changes must be documented. Use of a trusted third party for date and time setting will be easier for networked CDS systems. This subject is discussed in more detail in Chapter 16.

- *Backup and recovery SOP.* This SOP defines the secure backup and recovery process for the electronic records and software of the CDS. The procedure must include how to verify that backup worked correctly and describe what should be done if a backup has failed. Backup and recovery logs are maintained to demonstrate if backups have been successfully performed by usually checking the log files produced by the backup application or when data have been restored. Backups also produce electronic records (log files) but these should have a relatively short lifespan as the tape to which they relate to is usually routinely overwritten in the backup cycle. Backup and recovery is discussed in more detail in Chapter 16.

- *Defining E-records for the CDS.* A procedure for defining all electronic records (including metadata) within the CDS for systems for backup and recovery and well as inspections is required.[18] There should be a regular review of the defined electronic records for the system to ensure that the definition is still accurate and complete. Triggers for this review will be changes in working practices and reorganisations. Definition of the electronic records in a CDS is dependent on the way the system is used and Chapter 19 describes how to do this.

- *Record deletion.* Electronic records may only be deleted with justification and management authorisation required for deletion of records.[28]

- *Security and access control.* This procedure is needed to define how access is limited to authorised individuals. The scope should include both physical and logical access to the system. Define the access for each CDS user type and have an annual evaluation of the users and their associated access rights. Ensure that you include the vendor's service engineers in this list. This will document any changes in user access rights to system. If a user leaves or changes job their account needs to be disabled but do not delete it so that you do not create the same user identity for a different individual. There should be a cumulative record that indicates names of authorised personnel, titles and access privileges (both current and historical records). Prevent multiple logins by a single user wherever possible and users should be required to use a password-protected lockout on the lab client, and all logins must be recorded in the system audit trail.

- *Remote access.* If access to the system from outside of the organisation is allowed, then this SOP needs to define how it is achieved and the security necessary to access the system.

- *Uniqueness of electronic signatures.* For a CDS that uses electronic signatures typically involves user identities and passwords. There needs a way to ensure that the user ID is unique and that they are not re-used or re-allocated. The laboratory must ensure that users only use their own electronic signatures, not anyone else's even on their behalf, as that would constitute falsification. This

SOP can either standalone or it can be combined in the security and access control and/or use of electronic signature SOPs.

- *Verifying the identity of individuals.* Before a user can use electronic signatures, their identity needs to be verified and this procedure will describe how this will be done, *e.g.* birth certificate, identity card, passport, *etc.* This procedure may be integrated with a company's induction process but you must consider existing employees, new employees and change of name.

- *Forgotten password.* Describe the mechanism for resetting an account password if it has been forgotten. Under §11.300 (c)[17] there must be "stringent" controls, for example, a user has to show their identity card. There need to be records of when this SOP has been applied and identify the user who had forgotten their password.

- *Unauthorised access.* The CDS system administrator needs a procedure to describe how they will respond to attempted unauthorised access after notification by the system under §11.300 (d).[17] The CDS application should provide notification of attempted unauthorised access and should take preventative measures (*e.g.* lock a terminal after a specified number of failed attempts and/or retain card, *etc.*) then the system administrator will investigate and document the outcome.

CHAPTER 16

IT Support of the System

Networked CDS systems will require IT support and maintenance that will need to be formalised in a service level agreement (SLA). The IT group will also need to work in a compliant manner (*e.g.* written procedures with evidence that they have been followed) or the effort to validate the application will be wasted. Two of the key elements that will be included in an SLA are discussed here namely backup and recovery and time stamps.

16.1 What do the Regulators Want?

16.1.1 21 CFR 11

The impact of electronic record and electronic signature regulations[17] also means that data must be backed up effectively to avoid data loss as 21 CFR 11 has specific requirements that involve backup and recovery of your chromatographic data:

§11.10(c):

> *Protection of records to enable their accurate and ready retrieval throughout the records retention period.*

16.1.2 EU GMP Annex 11

Clause 13[27]:

> *Data should be secured by physical or electronic means against wilful or accidental damage, in accordance with item 4.9 of the Guide. Stored data should be checked for accessibility, durability and accuracy. If changes are proposed to the computer equipment or its programs, the above mentioned, checks should be performed at a frequency appropriate for the storage medium being used.*

Clause 14:

> *Data should be protected by backing-up at regular intervals. Back-up data should be stored as long as necessary at a separate and secure location.*

16.1.3 PIC/S Guidance

Section 19.1[31]:

> *The security of the system and security of the data is very important and the procedures and records pertaining to these aspects should be based on the IT policies of the regulated user and in conformance with the relevant regulatory requirements. The use of a computerised system does not reduce the requirements that would be expected for a manual system of data control and security. 'System owner's' responsibilities will include the management of access to their systems and for important systems the controls will be implemented through an Information Security Management System (ISMS).*

16.1.4 FDA GMP 21 CFR 211

Section 211.68(b)[12]:

> *...A backup file of data entered into the computer or related system shall be maintained except where certain data, such as calculations performed in connection with laboratory analysis, are eliminated by computerization or other automated processes ... Hard copy or alternative systems, such as duplicates, tapes, or microfilm, designed to assure that backup data are exact and complete and that it is secure from alteration, inadvertent erasures, or loss shall be maintained.*

16.1.5 483 Observations and Warning Letters

Furthermore, observations have been made during inspections such as:

- *"Failure to comply with network system backup procedures in that not all required backup procedures were documented as scheduled, showing lack of documented evidence that tape replacements were done."*
- *"There were no daily incremental backups; backups were performed every two weeks. If the system fails, data acquired between backups will be lost. The firm did not have contingency plans in case of a system failure."*

16.1.6 Regulatory Requirements Summary

IT components of the CDS must work in a compliant way including protecting the electronic records and data generated by the system. It also makes good business sense and is best IT practice.

16.2 Service Level Agreement

In the case of outsourcing or subcontracting the support for the servers that run networked chromatography data system software to the internal IT Department or an outsourced IT Department, a SLA should be written. This SLA should cover procedures such as:

- Backup and recovery
- Archive and restore
- Storage and long-term archive of data
- Disaster recovery

The content of a SLA is presented in Table 24. This SLA must cover the minimum service levels agreed together with performance metrics so that they can be monitored for effectiveness.

16.3 Backup and Recovery

The critical element in any SLA is the provision for backup and recovery. Without this in place and effective your data are at risk.

16.3.1 Business Rationale: How Important are Your Data?

The driver behind determining how often and how extensive your backup and recovery procedure is the importance of the data stored on the computerised system or network.

- How critical are the data? For critical data, the intervals between backups and the type of backups performed will be higher than low priority systems where backup may be made at a lower frequency.
- How often do the data change or new data are acquired? CDS that acquire new files regularly or where data are extensively manipulated will need more frequent backup than systems where change is less often.
- What speed of recovery is required? Can the system be restored within a working day with little impact or does it need to be restored within 4 h? This will affect the frequency and nature of the backup and recovery schemes as well as the linkage with database transaction logging.

All of these issues need to be considered when designing your backup and recovery process.

16.3.2 What is Backup and Recovery?

Backup and recovery is focused on storing and restoring system, application and user files.

- Storing means the copying from a source device, such as a disk on your CDS or network drive, to a target medium usually a magnetic tape.
- Restoring means copying from media containing stored files to the primary location of the files: usually a disk drive.

16.3.3 Roles and Responsibilities

There are two main roles involved with backup and recovery in a client–server CDS environment in most organisations:

- End-users
- IT Department personnel

Table 24 *Contents of a typical service level agreement*

Section	Detailed contents
Scope	CDS system subject to the SLA
	Specific exclusions
	Enhancements within scope
	Revision and distribution procedures
Roles and responsibilities	Laboratory responsibilities
	IT Department responsibilities
	Independent audit of IT Department
Definition of service	Service coverage
	Maintenance and housekeeping
	Continuity of service
	Access security
	Data security
	Regulatory compliance
	Network printing
Help desk	Response times
	Severity code
	Prioritisation of problems
	Status of problem
	Escalation
	Service contracts
Service reviews	Regular service reviews
	Quarterly reviews
Service levels	Service level targets – response times
	Service availability
	Service reliability
	Service report
	Service level targets review
Charging	Charging within the SLA
	Penalties
	Expenses
	Special charge considerations

End-users are responsible for backup and recovery as it is their system and their data. These facts are often overlooked when a backup schedule is developed. Responsibility can often be abrogated and a default schedule given that is not appropriate to the system or the data held on it. Users must be aware of their responsibilities in this area.

The IT Department will usually carry out the backup and recovery of data on behalf of the end-users. The schedule for the backups will be worked out in discussions with the end-users and this should be recorded in an SLA that outlines the roles and responsibilities of all parties and the schedule.

16.3.4 Hardware to help Data Security and Integrity

In many cases, data and application files are stored on a single computer disk. This means that there is a single point of failure that could mean the loss of your CDS files. This is why you should also consider the use of hardware options to consider improving data integrity and fault tolerance with the system. Usually, these hardware options are grouped under the acronym of RAID (Redundant Array of Inexpensive Disks). There are three options available for computers and servers commonly used to hold regulatory data:

- RAID Level 0: Data striping. This involves two separate drives where any data written to disk are broken into data blocks called stripes, these are written in sequence to both drives. The advantage of this configuration is speed, but the disadvantage is if a hard disk fails the stripe set will be lost and the data have to be restored from backup tape. Apart from the speed gains, there are few advantages to RAID 0. Therefore, for data security and integrity, one of the other two options should preferably be selected.
- RAID Level 1: Disk mirroring. Data are written to two disk drives that are configured identically. The difference between RAID 0 and 1 is that when a RAID 1 drive fails, the other drive contains an exact copy of the data and can be used immediately. However, to replace the defective drive the computer must be shut down for the defective disk to be replaced. There is also a single point of failure as there is usually a single disk controller for the two disk drives; however, if two controllers are used then this is termed disk duplexing. One potential disadvantage of RAID 1 (and this is exacerbated in the RAID 5 option) is that 10 Mb of data requires 20 Mb of disk space. However, given the cost of disk drives *versus* the value of the data stored on them, this is normally inconsequential.
- RAID Level 5: Fault tolerance, achieved by disk striping with parity, is essentially an extension of RAID 1. If a single drive fails then the data on it can be recovered from the other two by using the parity checksum information in the data stored on other disk drives in the RAID array. However, if two disks fail then its over to the tape backups to recover the data. To implement RAID 5, usually a minimum of three identical drives are needed and the operating system is set up to manage the disk array. In the normal operation of the computer, data will be copied across the disks with the parity checksums.

If a disk fails, some vendors offer a hot swap option where the old disk can be replaced with a new (but empty) disk and the data on the failed disk reconstructed using the parity checksums on the remaining two disks.

Normally RAID 1 or 5 would be used to store data. However, if your data are very critical, then you must use RAID 5 with the fault tolerance features to reduce data loss. If the computer application needs to operate with greater than 99% of uptime, then you need to consider additional hardware features such as dual processors and uninterruptible power supplies (UPS).

16.4 Options to Consider for Backup

There are a number of options that can be used in developing our backup strategy in addition to using hardware to mitigate the effects of a disk crash.

- *Full backup*. A regular backup of the system and is a complete copy of all system, application and user files on all disks to tape.
- *Incremental backup*. A regular, but partial copy of system, application and user files, identified by backup profile, to tape.
- *Differential backup*. A regular, but partial copy of files that have changed since the last normal backup.
- *Special backup*. Specifically requested copy of explicitly identified files to tape.

Most readers will be aware of the nature of a full backup of the system but will be less aware of what the differences are between incremental and differential backups. Let us look at a typical example of a backup.

- A full system backup is usually done at the end of the working week, say a Friday evening or over the weekend or when there are no users on the system. This will include all data and can also include the application and operating system, although the latter two are usually performed on a less frequent schedule, as there is less change. The OS and application software can be physically separated from the data using separate disk drives.
- During the week, incremental backups will be made on Monday, Tuesday, Wednesday and Thursday.
- Each incremental backup will contain the files that have changed since the last incremental or full backup.

Assume a system disk fails on Friday afternoon before the next full backup is scheduled. To recover the disk, the last full backup is restored and then the successive incremental tapes are also restored. Thus, to recover the disk back to Thursday evening, the full backup and four incremental backups need to be installed. A failure in one of the early incremental tapes will result in lost data even if the later backup tapes are perfect, thereby amplifying the impact of any data loss.

The differential backup, in contrast, contains all changes since the last full backup on the Friday night. Thus, the Thursday tape will contain all changes in files since Friday's full backup, *i.e.* the whole weeks work. After disk failure it only requires the full backup and the latest differential tape to bring the system back to Thursday evening. The differential tape backup will grow in size over a week, in contrast to an incremental tape that is relatively small in size. The greatest advantage is that only two tapes are required to recover data from a differential backup as opposed to a maximum of five with an incremental backup.

16.4.1 Main Backup Activities

The first part of the process is to schedule backups for the CDS system. How vital are the data stored on the CDS server? Only the users can tell you! If you work in IT, do not anticipate what the users may think or say. Get them to commit to the backup schedule themselves. The user department will usually be paying as is the case in most pharmaceutical companies today. The cost of IT services is recovered by cross charging for services provided to an individual system.

However, as protection of electronic records is mandated by 21 CFR 11 and the associated predicate rules, then if the CDS system is GXP classified, the data must be held for the records retention period as specified in the applicable predicate rule. For GMP, this is the expiry date of the batch plus a year. However, many organisations retain CDS data for longer than this and to protect against litigation many are intending to store data in perpetuity – meaning an upper limit has not been set. Criticality of the data will dictate the type of backup and the frequency so that any potential data loss is minimised.

16.4.2 Hot or Cold Backups?

There are two basic approaches to backup of computerised systems:

- Hot or on-line backup takes place whilst the system is still operating.
- Cold or off-line backup occurs when the system has been stopped and users have been logged off the system.

The cold backup is generally thought of as the safest type of backup as the hot backup requires the system to be buffered while the backup occurs and the system updated when complete. The option you select is up to you depending on the use of the system and the value of data.

For instance, if you have a data system that is required to be available 24/7 (24 h per day and 7 days per week) then a hot backup of the system would be required. Alternatively, if the system was only required to be available 95% of the time, then a cold or off-line backup approach could be devised where the system would be backed up when there were no users on the system.

16.4.3 Cold Backups

A cold backup is scheduled out of working hours, say at 2 a.m. in the morning and is performed automatically with the logs of the activity reviewed the next morning to confirm that all has gone well. The process is as follows:

- *Remove users from system.* If you need to backup the system during normal working hours warn the users of impending downtime, then ensure that all users are logged off and have their files closed. The system manager can disable further user access and also making sure the appropriate processes for backup are running.

For normal hours or out of hours backup the common process is:

- *Copy files to media.* Using a software tool provided explicitly for the purpose of backups (either with the operating system or purchased specifically for the purpose), files are copied from their primary location on disk to the backup media, usually tape. Typical backup applications are Backup-Exec and ArcServe.
- *Verify readability of backed up files.* You are placing your trust and your data in the hands of a magnetic medium, therefore the quality of the backup is essential. Verification confirms that the files have been copied to the tape correctly and that the tape can be read again. Verifying readability of files backed up can include a comparison of the files on the backup media with the originals on disk and this gives the best confirmation, but this can take time. Verification is important, as backup is a vital component of data security and integrity, for off-line backups this step usually occurs before users are allowed access to their applications following the backup.
- *Allow user access to system.* If the backup is verified as readable, the system can be returned to normal operations.

What happens if a backup fails? Unsuccessful backups must either be rescheduled for the next backup window or a backup is performed as soon as the problem is known. However, a cold backup requires that users must be logged off the system and this may result in down time of the CDS. The system owner has to balance the potential loss of data against system downtime, in this instance it may be worthwhile considering a hot backup schedule.

16.4.4 Hot Backups

Hot backups require a fast tape system to transfer the data whilst the system is operational. There may be a slight but noticeable degradation of performance but the availability of the system overrides this issue.

One way to overcome this is to have a second disk that is empty and the same size as the data disk on the server. Transfer the data from the operational disk to the empty disk, this is a relatively quick operation as the original disk reading, transfer *via* the internal bus and disk writing is far quicker than the disk to tape transfer. When complete, the image of the data on the second disk is backed up as if it were off-line. When the backup is complete and has been verified, the data on the second disk can be deleted. The disadvantage is the cost of the additional disk and any associated service costs from the IT Department. However, in my view, the benefits of this approach greatly outweigh the disadvantages.

16.4.5 Media Management

Media are typically magnetic tapes of suitable size to backup the disk with a single tape. Media management is defined as the activities necessary to ensure that backups and restores have reliable media, where they need it and when they need it. This can take a number of forms such as:

- *Media identification.* Ensuring that all tapes are uniquely identified with a number, colour and if an automated robot is used then a bar code can be used along with human readable information for any manual storage.
- *Media rotation.* Regular cycling of media used for backups is required which includes replenishing supply and disposing of unreliable media. Media are considered unreliable when they have been used beyond their normal supported life, when they have been found to be unreadable, when they have flaws making them unusable or an error is reported during backup. This is critical: do not think you can save money by reusing suspect media as you may pay a much higher price in the long run through data loss.
- *Logical media library.* A catalogue of the backup media with retrieval index, contents and location for each system.
- *Media audit.* Verification that media can be found at the location specified, are readable and contain the data specified and are listed in the logical media library. Audits can be scheduled at regular frequencies to confirm that there would be no problem locating the appropriate tapes.
- *Dual locations.* Once every 2 weeks or a month, full backup duplicate tapes are made and they are stored in a separate location either on the site or off-site as a disaster recovery measure.
- *Manage backup media generations.* Depending on retention policies, determine which backup generations can be reused and which must be saved. For example, if retention indicates that 3 months of the first full backup of each month are to be saved, and it is the middle of September, the media from June can be reused.
- *Determine additional needs for new media.* If media use is increasing due to higher volumes of data being backed up, more frequent backups, or other changes in the backup profile, there may be a need for additional media. Plan proactively for this rather than run out of tapes and have no cover.

16.4.6 Restoring Data from Tape

Despite all the efforts of designing fault tolerant hardware, there will be a time during the operation of any system that anything from a single file to a whole disk will be need to be restored. There is where the appropriate tape is invaluable, assuming the backup has been done correctly and tape can be read.

- *Request media from library.* You will need to identify the tape or tapes that the data are on and bring them to the tape unit for the system. A restore request will usually indicate the file(s) to be restored and the date of the backup. Thus, the media request resulting from this process will identify the media to be used.

- *Execute and verify restore*. Using the correct tape, identified through your super effective library catalogue that you validated before it became operational. The file or data requested is restored to your system. Of course, we will not forget to verify that the recovery has worked. Database recoveries can be a little more complex than simple files. For example, a database recovery might entail recovering the log files (*e.g.* redo logs) following a restore.
- *Return media*. Tape(s) are returned to the library.

16.5 Time and Date Stamps

To ensure trustworthiness and reliability of electronic records, the accuracy of the time stamp applied when a record is created, modified or deleted is vital. In a paper environment, the time sequence of events is laid out in the linear sequence of how a laboratory notebook is completed but this is not true of the electronic world of a CDS. Therefore, the means of setting and checking the accuracy of the time and date stamp is important.

16.5.1 FDA Guidance on Time Stamps

The FDA issued in March 2002 a draft guidance for industry on time stamps[24] and then withdrew it in February 2003.[18] Notwithstanding the withdrawal, this guidance[24] has good useful information and the key points and discussions for CDS applications are:

- The time stamp needs to be accurate to "within a minute"; this is best interpreted as ±1 min.
- The location of the time stamp needs to be defined. This is a change in the FDA's approach as the preamble to Part 11[17] that required the time stamp to be local to the user. For a system working on one site this is not an issue. However, for CDS that is operated across countries and time zones this is critical. What happens if the network is down? How will local laboratories cope until the system is restored? What happens to the time stamp in these cases? Knowing how the system handles times is the key issue here. In the final version of the Part 11 Scope and Application guidance[18] there is a footnote that confirms that the FDA still want the location of the time stamp to be defined.
- The guidance makes no reference to summer and winter daylight savings or to leap years. Typically, these changes will be managed by the operating system which requires an accurate time and date stamp setting and you may want to check that these have occurred the next working day after the event.

16.5.2 Time Stamps for Standalone CDS Systems

Time stamps for standalone CDS systems are probably best set and maintained manually. Although there are technology options they will be relatively expensive to implement on a large number of individual workstations. You will need a manual procedure with records to show that the time stamp has been checked and maintained against a standard time source, *e.g.* speaking clock. How often you will need to check the time stamp will vary. Start on a monthly basis for the first 6 months then review how accurate the time stamp has been over this period and if you have needed to adjust the clock. You can then lessen the frequency if necessary on the basis of experience. You will need to maintain a log of who checked the computer clock, the computer time, the standard time and if any adjustment was made.

Being realistic and practical, this is a waste of resources especially when coupled with the need to manually backup individual systems. Ideally, all CDS systems should be networked to ensure that time stamps are maintained from a single source within the network.

16.5.3 Time Stamps for Networked CDS Systems

This is by far the easier and preferred option for setting and maintaining time stamps. Within the network a server provides the time stamps for all other servers and workstations. As a minimum, each time a user logs on to the network from a workstation the time stamp is updated automatically. The operating system can also be set up to update the time stamp during the time that a workstation is logged on, *e.g.* hourly if required.

Trusted time sources are available to automatically check and correct the network clock; this can occur in a number of ways:

- Network Time Protocol (NTP) where the time-server accesses a time source on the Internet.
- Time sources from the National Observatories of some countries, *e.g.* US Naval Observatory has atomic clocks in Denver and Washington and there is the Frankfurt atomic clock for Europe and the UK has a source at Rugby.
- Global Positioning Satellite (GPS) system also can provide a Universal Coordinated Time signal that can be used with the appropriate equipment to provide a check of server time.

The time signal is then interpreted by the operating system to the local time zone and any daylight saving that has been implemented.

Checks of the time setting should be limited in these cases to seeing if daylight savings if implemented has occurred every spring and autumn and if leap years have been incorporated every 4 years.

CHAPTER 17

System Description

The system description is an approved and controlled document that outlines the main elements of the CDS system, what it does and who uses it.

17.1 What do the Regulators Want?

Why do we need a system description? The simplest answer is that it is a regulatory requirement! In European Union GMP[27], OECD GLP Guidance[28] and the new PIC/S Guidance for Inspectors on Computerised Systems in GXP Environments[31] there are the following requirements and statements on a system description:

17.1.1 EU GMP Annex 11

Clause 4[27]:

A written detailed description of the system should be produced (including diagrams as appropriate) and kept up to date. It should describe the principles, objectives, security measures and scope of the system and the main features of the way in which the computer is used and how it interacts with other systems and procedures.

17.1.2 OECD Application of GLP Principles to Computerised Systems

Section 8.5[28]

For each application there should be documentation fully describing:

- *The name of the application software or identification code and a detailed and clear description of the purpose of the application.*
- *The hardware (with model numbers) on which the application software operates.*
- *The operating system and other system software (e.g. tools) used in conjunction with the application.*
- *The application programming language(s) and/or database tools used.*
- *The major functions performed by the application.*
- *An overview of the type and flow of data/database design associated with the application.*

- *File structures, error and alarm messages, and algorithms associated with the application.*
- *The application software components with version numbers.*
- *Configuration and communication links among application modules and to equipment and other systems.*

17.1.3 PIC/S Guidance

Section 23.13[31]:

> *The lack of a written detailed description of each system, (kept up to date with controls over changes), its functions, security and interactions (A11.4); a lack of evidence for the quality assurance of the software development process (A11.5), coupled with a lack of adequate validation evidence to support the use of GMP related automated systems may very well be either a critical or a major deficiency. The ranking will depend on the inspector's risk assessment judgement for particular cases. (NB. Since 1983, the GMPs have called for validated electronic data-processing systems and since 1992 for the validation of all GMP related computer systems).*

Note the numbers in brackets, *e.g.* A11.4, refer to EU GMP Annex 11 clauses; the first GMP reference is to an FDA compliance policy guide 7132a.08[66] and the second is to European Union GMP[27].

17.1.4 Regulatory Requirements Summary

So you need a system description and it needs to be an approved and controlled document. Furthermore, the lack of this document can give rise to a critical or major observation. The main purpose of a system description is to give an overview of the CDS for an inspector but it can also be useful within the organisation for introducing new staff to the system.

17.2 Turning Regulations into Practice

How do we write a system description? The easiest approach is to use the regulations to give you the format for the document. If you use the framework described in the OECD GLP consensus document your system description can be based on this[28].

17.2.1 Single Document or Multiple Documents?

Before you begin, think the process through. You do not have to cram all the information into a single document. For instance, the hardware and operating system information required will typically be found in the configuration management records for the CDS or the current version of a technical architecture document. Instead of duplicating the information in a second document and have to update two documents when a change is made, simply cross-reference the configuration management records in the system description. Furthermore, the system description should not be very long document – it is a summary document: an overview not a specification.

Table 25 *Contents of a CDS system description*

Section	Description of contents
Title and approvals	• Title of the document including reference to the name and version number of the CDS application • Approval and release of the document
Introduction	• What is the purpose of the system • Name of the CDS application and supplier • Departments or laboratories using the system
Referenced documents	• List of documents for key information e.g. configuration records for the system security
System scope	• Major functions of the system: diagrams of the CDS process workflow • How data are stored: database or file structures • Diagram showing any interfaced applications e.g. LIMS with the main data transferred between them
Definition of electronic records	• Defining the electronic records for the system as required by the FDA Guidance on Scope and Application[18] and detailed in Chapter 19

Therefore, put a referenced documents section into the system description to reduce the amount of detail required.

17.2.2 Outline for a System Description

The content of a system description is outlined in Table 25. We will discuss the main sections of the document in the text as well. Note that many of the OECD requirements outlined in Section 17.1.2 are intended for a system developed in-house therefore, the need to document the programming language, errors and alarm messages, algorithms, database design or tools used is not usually applicable for a CDS.

17.2.3 Keeping Current: Updating the System Description

The system description, like the URS, is a living document as noted in the PIC/S Guidance[31] and EU GMP regulations[27] and it needs to be reviewed and updated every time a new version of the CDS is implemented as well as when a company reorganises or merges. In either instance, there is a situation when the way the system is used, the scope and purpose or the electronic records generated by the CDS may change. Therefore, the document needs to be reviewed and reissued to reflect this.

If new clients or instruments are added, do you want to update both the system description and the configuration management records? No. Think the process through – the system description document is an overview perhaps with a summary diagram of the system. The configuration management records provide the detail.

This approach means that the system description will only be updated for key events when the software is upgraded from one version to the next.

When considering the first version of the system description, keep in mind that the document will be change and you should write it with ease of update in mind. Therefore, something as simple as the version number of the CDS just refer to it once and then it is easy to update. For example, in Table 25, there is just a single mention of the version number in the title of the document. The reason is that it is highly visible on the title page and easy to update when necessary.

17.3 Key Sections of the System Description

17.3.1 Introduction

This should be a short section that presents the essence of the CDS as a simple set of statements, for example:

> *The SuperChrom application supplied by a MegaSoft is a chromatography data system used to acquire and collect data from gas and or liquid chromatographs in support of R&D, primary or secondary manufacturing departments. It is operational in the Analytical Development, Bulk Chemicals and Finished Goods Quality Control Departments of the company.*

The aim is to present information in an accurate and precise manner. Remember, who the readership of this document will be. It is important at this stage also to manage expectations. If the CDS is supporting more that one site separated geographically either within a country or in separate countries, it needs to be mentioned here in the introduction. Therefore, an additional sentence needs to be added as follows:

> *The CDS supports three sites at London, Birmingham and Manchester connected via the corporate WAN with the server located in London.*

17.3.2 System Scope

The system scope covers the major functions of the system, data storage and any interfaced applications to the CDS. Diagrams can be of great use in helping to present a concise overview of the system. Remember that there is also the URS and other documents that are available to describe the functions in more detail.

For example, a diagram similar to that in Figure 5 in Chapter 2 would give the overall workflow of the system with the major functions. This should be accompanied by some explanatory text which would be sufficient detail for the main functions of the system such as sequence file, method files, instrument control, data acquisition, integration, calibration, calculation of results and approval of them (either manually or by electronic signatures). Further discussion of the functions can be eliminated by cross-reference to the URS.

Diagrams are also useful to describe the interface with other applications. Figure 37 shows a LIMS interfaced with a CDS. The data flows between the two applications are shown at a high level only. Sample numbers and weights are transferred down to the CDS to be incorporated into sequence files for a single

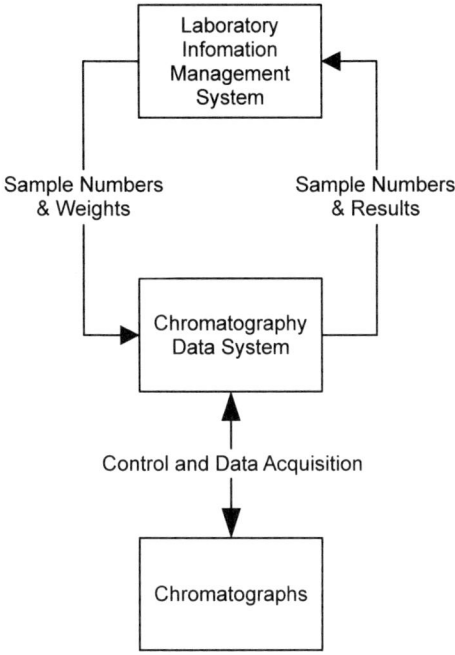

Figure 37 *System description diagram showing a LIMS interfaced to a CDS*

chromatographic analysis. After the analysis and data interpretation, the calculated results associated with the appropriate sample numbers are transferred to the LIMS for reporting.

This is one example of many possibilities of interfacing the two applications. In another laboratory, the peak areas could be transferred up to the LIMS for centralised analysis and calculation of final results rather than the CDS doing the work.

17.3.3 Definition of Electronic Records

As outlined in the Part 11 Scope and Application guidance[18] the FDA recommend that this is defined and documented. One place that this can occur is the system description if desired. The detailed discussion of what constitute electronic records for a system is covered in more detail in Chapter 19 and will not be discussed further here.

CHAPTER 18

Validation Summary Report

The validation summary report brings together all of the documentation collected throughout the whole of the life cycle and presents a recommendation for management approval when the system is validated. The document should also contain a formal release statement to allow the system to be used for regulated work.

18.1 What do the Regulators Want?

18.1.1 PIC/S Guidance

Section[31]:

> *Inspectors should review the firm's Validation Summary Report*, (VSR) for the selected system and refer as necessary to the System Acceptance Test Specification and lower level documents. They should look for evidence that the qualification testing has been linked with the relevant specification's acceptance criteria, viz:*
>
> - *PQ versus URS*
> - *Supplier audit reports*
> - *Validation plans*
>
> **VSR = A best practice high level report, summarising the validation exercise, results and conclusions, linking via cross referencing to lower level project records, detailed reports and protocols. This is useful for briefing both senior managers, in regulated user organisations and for reference by auditors/inspectors.*

18.1.2 General Principles of Software Validation

Section 5.2.6: User Site Testing[29]:

> *User site testing should follow a pre-defined written plan with a formal summary of testing and a record of formal acceptance.*

Section 6.2: Defined User Requirements[29]:

> *The device manufacturer should have documentation including: defined user requirements; validation protocol used; acceptance criteria; test cases and results; and a validation summary that objectively confirms that the software is validated for its intended use.*

18.1.3 Regulatory Requirements Summary

The aim of the validation report is a summary document not a full-length novel of Nobel Prize for Literature proportions. As the footnote above in the PIC/S guide says it is a summary with cross-references to the documents detailing the actual work, this allows a quick and effective means of issuing the VSR.

18.2 Content of the Validation Summary Report

One outline of a summary report, based on the IEEE standard 1012,[51] is presented in Table 26. The emphasis is on using a summary report as a rapid and efficient means of presenting results as the detail is contained in the other documentation in the validation package.

The whole life cycle and the documentation resulting from the activities can be reported in this way, however if additional detail is required it can be cross-referenced within the report. There should also be a section at the front of the report where management authorise the operational use of the system within a regulated environment.

18.3 Writing the Validation Summary Report

The VSR collates together all of the information and activities collected and undertaken throughout the whole of the life cycle and presents a recommendation for management approval that the system is validated and should be released for operational use. The emphasis is on using a summary report as a rapid and efficient means of presenting results as the detail is contained in the other documentation in the validation package.

The contents of a VSR are shown in Table 26. Each of the major phases of the system development life cycle is represented. This is based on an IEEE standard 1012 for validation and verification plans. The major change from the standard is the addition of a section called validation documentation (this can also be called validation package, dossier or registry). Essentially it is the list of all the documents that support the CDS validation.

18.3.1 How to Summarise the Work

The issue is how to summarise a phase of the life cycle. Here is an illustrative example of how a URS could be summarised:

> *A User Requirements Specification (URS) was drafted and revised between September and November 2004. Version 1.0 of the document was approved in early December 2004. This specified the intended functions that the system would undertake as well as the capacities of several functions and system support requirements. Each requirement is uniquely numbered as well as prioritized as either mandatory or desirable.*

The text simply needs to give an overview of the work done; if an inspector wants to see more all they need to do is request the URS.

Table 26 *Contents of the validation summary report – adapted from IEEE Standard 1012*[51]

Validation statement and release for operational use

Introduction

Purpose

Objective and scope of validation efforts

Life cycle activities and documented evidence for system validation

- Evaluation of the requirements phase
  Validation plan
  User requirements specification
  Risk analysis and traceability matrix
  Summary of the requirements phase

- Evaluation of the vendor's design, build and test phase
  Vendor audit to cover design, programming and developer testing
  Vendor product certificates
  Summary of implementation phase

- Evaluation of the qualification phase
  IQ and OQ of the components and software
  Documentation of system and software configuration
  Performance qualification test plan and test scripts
  Execution of the PQ test scripts
  PQ test execution notes
  Summary of PQ anomalies and their resolution
  Summary of qualification phase

- Evaluation of training, documentation and procedures
  User training
  User procedures
  System (IT) procedures
  Vendor documentation including on-line help
  System description
  Summary of training, documentation and procedures

- Deviations from validation plan and their impact on quality
  Validation documentation/package

- List of all the documents that make up the validation of the CDS

18.3.2 How to Summarise PQ Testing

From Chapter 14 you will remember the PQ test plan as a document that contains the overall approach to the PQ or end-user testing. In this document, the test scripts and the outline testing approach are presented. A quick way to present the results of the PQ testing is to copy and paste the same tables from the PQ plan into the appropriate section of the VSR and add in the left-hand column a summary of the testing outcome; this is shown in Table 27. The advantage of this approach is that it

Table 27 *Summary of PQ testing in the validation summary report*

Chromatography data system requirements to test

Test script identifier	Test outline	URS requirements tested	
TS01: Security and Access Control	• Logical security for access to the system *via* the clients will be tested	4.3.08 4.4.01	4.6.01 4.6.02
	• Access by the different levels of user to specific functions will be tested, *e.g.* administrator, analyst, operator and end-user	4.4.02 4.4.04 4.5.01	4.6.03 4.6.04 4.6.05
	• Test groups across geographical locations: *i.e.* users in same group but located in both sites can see same data	4.5.02 4.5.05 4.5.06	4.6.06 4.6.10 5.3.02
	• Appropriate users are able to create and modify methods	4.5.09	5.3.03
Tests executed with expected results	• A user's access rights can be reconfigured by the system controller	4.5.10 4.5.11	5.3.04
PASS	• Automatic log-out of the system after approximately 3 min when in electronic signature sign-off	4.5.17	

is quick and simple to write but also means that the requirements are traceable from the URS to the VSR.

18.3.3 PQ Test Execution Notes

Let us face it, you are not going to write the PQ test scripts perfectly as discussed in Chapter 14 and there will usually be test execution notes written up in the course of the execution. Most will not be major and they will remain in the test script. However, the ones that are major should be documented and discussed in the VSR.

For example, the following issues, in my view, should be noted in the VSR and discussed:

- A manual calculation formula that is used to check a data system calculation is wrong is noted and changed during the execution.
- A method that was allocated to a test script is not used and another substituted.
- Test incidents or software anomalies that impact the quality of data generated or operation of the system.

18.3.4 Deviations from the Plan

Before you circulate the first draft of the VSR for review, make sure you have gone back to the validation plan and PQ test plan and read them both thoroughly. The validation plan is the documented evidence of intent and the VSR is the documented evidence of what was actually done. The PQ test plan covers what is usually the greatest portion of the validation effort, the end-user testing. The deviations from plan section discus any departures (planned or unplanned) from what was originally described in the validation or PQ test plans along with a discussion of their potential impact on system quality.

For example, the PQ test plan may state that you will have a certain number of PQ test scripts to write and execute. During the writing of the test scripts you could have a situation where a better and more elegant way of testing functions covered by two test scripts is identified and a single test script would result not two.

You have two options: reissue the PQ test plan with the modifications or issue a file note or equivalent that is approved by the system owner and QA that two test scripts will become one. If you choose the second option you can use the deviations section in the VSR to note and discuss this approach. This is a planned deviation and still tests the same functions and will not have any impact on the overall validation.

18.3.5 Validation Package

There should be a cross reference within the VSR to the validation documentation or document number so that the document could be easily retrieved if required. Some VSRs I have reviewed have been simply a list of documents produced with a release statement and others have had descriptions of the work carried out in each phase of the project. Whatever your approach, it is important to bear in mind who will be reading this document: QA and regulatory inspectors. It could be one of the first documents requested in an inspection, and therefore you need to use it to generate regulatory confidence. In my view a shopping list is not the best way to do this. A good VSR tells the story of your CDS validation and needs to be succinct but generate confidence. Therefore, spend a little more time on the report and a better document will result.

18.3.6 Releasing the System

The release statement that the system is fit for operational use should be completed by the validation team and authorised by the system owner and quality assurance. It is a simple statement that the system is released for use in a GXP environment. However, there may be some conditions attached depending on the results from the validation effort. For example, a calculation provided by the system may be mathematically incorrect or the system may state that it is one calculation but the formula actually used is different - do not laugh it has actually happened.[58] Therefore, the system may have caveats for some functions to be under procedural control or may not to be used for GXP work. This should be noted in the release statement.

If this issue is resolved later with a service pack or new version of the software, the operational release of the revalidated system can dispense with the constraint.

18.3.7 Going Live! Sit Back and Relax?

You may think all the hard work to validate the system is over now you have gone live, but you have just finished the easy part of the validation of your CDS. The most difficult part of validation is now before you. Maintaining the system in a controlled and validated state during the whole of the operational phase some 5–10 years is the hard part of the job.

CHAPTER 19

Defining Electronic Records for a CDS

What are the electronic records generated and maintained in a chromatography data systems? Why do we need to change our approach to managing electronic records compared with paper ones? The discussion in this chapter focuses on the definition of electronic records produced by a CDS and this is influenced by the way you use the system as defined in your URS.

It is debatable if this chapter should be in the section before the operational release of the system or after. I have chosen to place it in the operational section as the way the system will be used will usually change over time and therefore the electronic records generated by the system will reflect this change.

19.1 What do the Regulators Want?

19.1.1 21 CFR 11

Electronic records definition:

> *any combination of text, graphics, data, audio, pictorial, or other information representation in digital form that is created, modified, maintained, archived, retrieved, or distributed by a computer system*[17]

19.1.2 FDA Part 11 Scope and Application Guidance

Part 2 Definition of Part 11 Records[18]:

- *Records that are required to be maintained under predicate rule requirements and that are maintained in electronic format in place of paper format. ... We recommend that you determine, based on the predicate rules, whether specific records are part 11 records. We recommend that you document such decisions.*
- *Accordingly, we recommend that, for each record required to be maintained under predicate rules, you determine in advance whether you plan to rely on the electronic record or paper record to perform regulated activities. We*

recommend that you document this decision (e.g., in a Standard Operating Procedure (SOP) or specification document).

19.1.3 Regulatory Requirements Summary

The definition of electronic records is very wide-ranging and far more extensive than paper; therefore, what constitute electronic records in your CDS must be documented so that they can be managed and maintained.

19.2 Literature Contributions to the E-Records Debate

19.2.1 Furman, Tetzlaff and Layloff (1994)

Although this chapter focussed on the qualification of HPLC equipment, there was a discussion of electronic raw data that drew on the updated definition by Furman *et al.*[60] that "raw data were all those that could be saved and accessed later." The authors pointed out that chromatographers should consider the following scenario: two small peaks are detected eluting ahead of a peak of interest which are ignored in the current method. Later evidence finds that these are minor compounds, are toxic and the organisation needs to know how much of these compounds was present in all batches of material analysed in the past year. Furman *et al.* then posed the question of which is preferable: reanalysing all batches of material or retrieving the old files of raw data and reintegrating them?

Furman's bottom line was very conservative: save all raw data including the chromatographic time slice data files and the methods used to acquire the data, fit baselines and calculate results.[60]

19.2.2 BARQA

The second main reference was the paper[67] by BARQA (British Association of Research Quality Assurance) that outlined that raw data mainly within the context of GLP had four key factors:

- Original records or records of original observations.
- Recorded directly, promptly, accurately, legibly and indelibly with observer identified. Raw data generated by direct input should be identified by the individual responsible for data entry.
- Changes to raw data should not obscure the original.
- There is a wide range of media that could be defined as raw data.

The advantages of defining electronic raw data are that storage is compact and efficient and, like the argument from Furman, the data are available for further processing and analysis. Data can be copied with ease to provide duplicates (back-up) to give increased security against loss or damage, providing it is documented and the copying is authenticated.

19.3 Non-compliant Working Practices

Poor practices still occur in many chromatography laboratories, as the message to retain and preserve electronic data still has to sink in. Here is just a sampler of those working practices – of course, these never occur in your organisation, do they?

Still deleting files? For instance, when working with stand-alone PCs that run CDS applications, when the hard disk gets full, the easiest way to resolve the problem is to delete those pesky files that are clogging up the hard drive. A few commands or a couple of mouse clicks and your data storage space problems are solved. After all, we still have the printed paper to fall back on, don't we?

Overwriting method files? One of the features of some data systems is the ability to overwrite method files. Occasionally, you may get a message that the system is going to do this or more likely the system just does it. Lovely feature, but how can you retrieve the same method that was used on the same day you acquired a particular set of data files? Not a hope.

No audit trail? You will need to have an audit trail to help ensure that your records are acquired, manipulated and reported correctly and with no falsification of results. The lack of one is a major requirement for compliance. Although the FDA guidance on Part 11 Scope and Application[18] allows enforcement discretion for Part 11 requirements for audit trails; it is a double-edged sword. The predicate rules require records of changes made to data and GMP requires "complete data"[12] for laboratory records. Therefore, for a multi-user client system at a minimum and ideally all CDS systems, an electronic audit trail is essential for cost-effective regulatory compliance rather than any of the other options mentioned in the Part 11 guidance.[18]

19.4 Culture Shock: Changing from Paper to Electronic Records

We all need to change our approach and mindset as we move from paper raw data to electronic records. Consider the paper record first. We have a pile of printed paper. Depending on your working practices, there may be a record in the pile of the:

- sequence of injections
- instrument control parameters
- method used to acquire and process the data
- interpreted chromatograms with or without baselines drawn on them
- calibration method
- system suitability samples and calculations
- peak areas or heights
- any post-run calculations, *e.g.* dilutions, mass or volume adjustments, *etc.*
- analyte amount or area normalisation results

You will pick up this paper from the printer, check it and then file it. If anyone wants to check it at some later date, some poor individual goes on an Indiana Jones expedition into the bowels of your organisation's archive, moving cobwebs out of the way and digs out your data package. You are dealing with a tangible and physical medium that we all know and have used all our lives.

We must have a different mindset when we move to electronic records. However, the durability and robustness of the storage media change dramatically when moving from paper to electronic media. Magnetic media are erasable and friable.

Electronic records must be considered in a much wider context than their paper counterparts. Note, however, that the data retrieved tend to be similar to the list above *but* you will need the actual electronic version of the method file used to acquire the data not a generic version. Your working practices must change to reflect the new electronic age. Similarly, the working practices and features of commercial chromatography data systems must also change. For instance, overwriting of files must not be allowed. However, before we can proceed much further, we have to consider the term "meta-data".

19.4.1 Meta-Data

The electronic records debate about what constitutes an electronic record is widening as the impact of 21 CFR 11 is understood in more detail. The term *meta-data* is used to describe the additional files needed to process and then reprocess the data. Unfortunately, the term *meta-data* is not defined in any regulations or guidelines and, in my view, this can cause confusion especially as it is a computer term that tends to be unfamiliar to chromatographers and other scientists.

However, to help understand this and the wider context that electronic record keeping brings, recall the definition of Furman *et al.*[60] when they wrote all data should be saved including fitting peaks, *etc.* We must go further along these lines to understand what we must now save as electronic records for a chromatography data system. I will describe this in terms that will be familiar to all of us in the chromatography laboratory.

19.4.2 Back in the Lab

Let us look in more detail at what we can define as electronic records for an individual chromatographic run. The following descriptions you will have to look at and modify for the specific CDS that you are using within your individual laboratories. Some functions may be called something else or the system works slightly differently.

I will assume that you have an automated chromatograph controlled by a CDS. The system is an isocratic HPLC with a variable wavelength UV detector commonly operated in many chromatography laboratories worldwide. For those readers that only have gas chromatographs, the principles will be the same you will need to adjust the approach for the differences between LC and GC. The basic files that should constitute electronic records are shown in Figure 38, with the exception being the method of storing the data files that I will describe now.

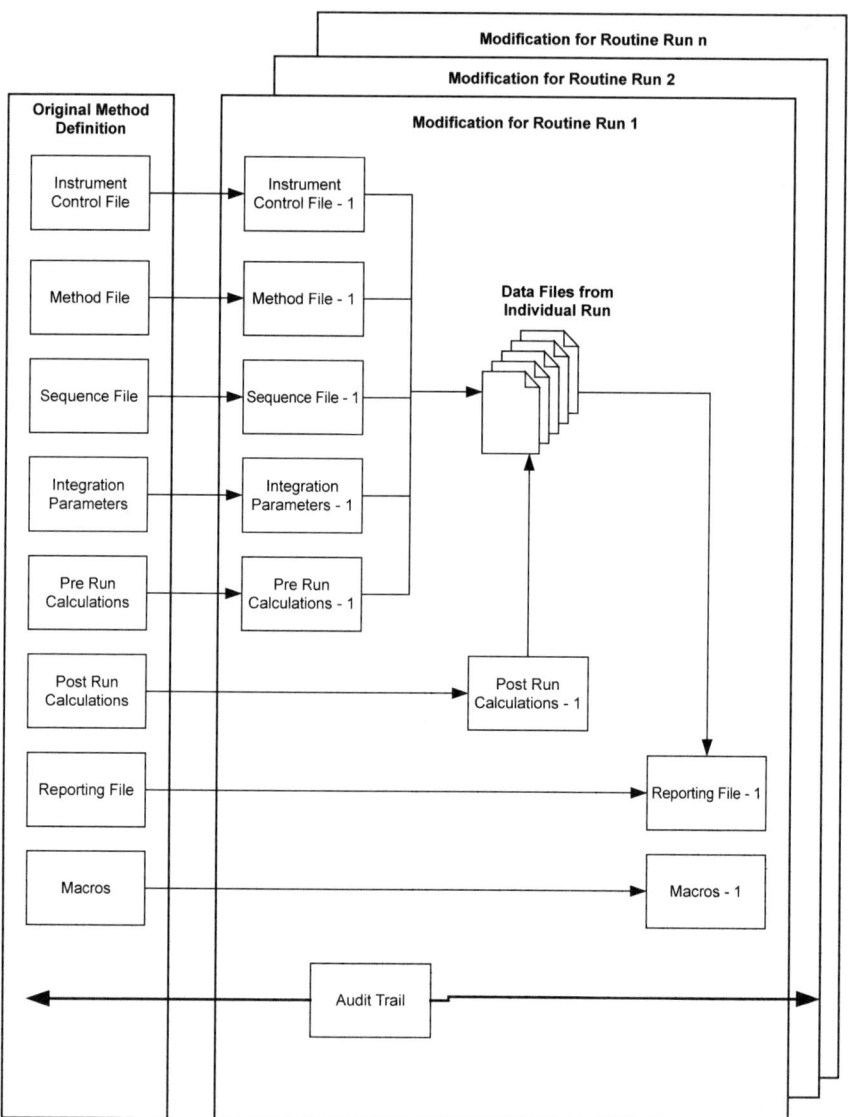

Figure 38   *Basic electronic records for a chromatography data system*

19.4.3 Data Organisation

Before you embark on collecting data, you will need to consider how you are going to store them. Usually, this will involve using a directory structure defined using the operating system or a database (dependent on the CDS you use). Either way, you will need to ensure the storage is set up for easy archive to remove *all* the electronic records associated with your analysis from the system. Currently, some data systems may aid you in this process but most will not, so plan and consider all the aspects.

Organising the data will be a key aspect here as you will need to ensure the directory or data table you want is defined to locate the files generated from the chromatographic run. The larger the work package, the more you may have to put into manage all data and the need to name your files appropriately.

Naming conventions for data files, methods, sequences and reporting methods may also be important and should not be underestimated. Get this right now – otherwise you will be in trouble later in the archiving phase of your work and an army of Indiana Jones's will not dig you out of this particular hole.

19.4.4 Instrument Control and Calibration

Again, this is dependent on the type and complexity of data system and the instrumentation interfaced to it. As shown on the right-hand side of Figure 39, a chromatograph controlled by the data system will usually have an electronic file for instrument control within the data system that is either part of the method or linked to it in some way. If the CDS can be used to check that the instrument is working correctly, then the data files for each analytical run need to be saved as part of the electronic record. For some mass spectrometers used as GC or LC detectors, the data system can control and calibrate the instrument. Data files from calibration runs should also be available to demonstrate that the instrument worked acceptably before an analytical run was started.

However, there is also the relatively common situation where an instrument is locally controlled by the vendor's data system; this is shown on the left-hand side of Figure 39. The central data system acquires and processes the chromatographic

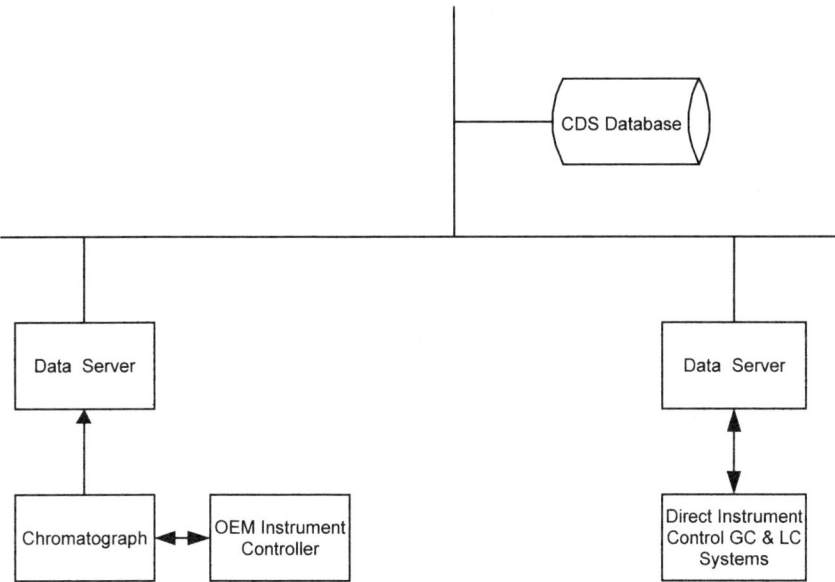

Figure 39 *System architecture and its impact on definition of electronic records*

data (analytical runs) but the instrument vendor's data system controls the chromatograph. Typically, the instrument control files from the Original Equipment Manufacturer's (OEM) data system are incompatible with the central CDS used for data processing. This can present a problem, but here you can argue that a paper printout of the daily instrument conditions can substitute for the electronic records under the Part 11 Scope and Application guidance.[18]

19.4.5 Setting Up an Analytical Run

You will have a method file that will describe how the instrument will be controlled and how the data acquired over the time of the analytical run. There will also be an associated file detailing the sequence of samples to be injected into the chromatograph from the autosampler or injector and an integration file that will determine how the data system will interpret the files and place baselines and carry out any post-run calculations.

These files will be stored on the data system as a master set and used for each analytical run. However, what happens in practice is that you will fine-tune individual files on the basis of the chromatographic parameters found each day you run the method. Reasons for change may be one or more of the following examples:

- retention times will vary slightly depending on the condition of the column
- how well a chromatographer has made up the mobile phase
- ambient temperature of the laboratory (if the column is not temperature controlled)
- robustness of the method
- pump seals and condition of the check valves

Each version of your method, sequence file and integration file that is used with a specific analytical run needs to be saved. Some data systems are good about doing this, others are less so and only the latest version is saved.

Therefore, additional safeguards need to be considered within the data system and reinforced by training and documented working practices. Special attention needs to be paid to version control of these files with no overwriting allowed. These features on your data system will need to be linked to the security profiles of your users. Who is allowed to create, modify and delete files?

The sequence file will contain the run sequence of samples, quality controls, blanks and standards and the number of replicate injections to be made for each sample. There may be some pre- or post-run calculations required, *e.g.* entry of sample weights, dilution factors. There may be a download of information from a spreadsheet or LIMS containing sample identity. This file is very important and must be saved; otherwise, reprocessing is not feasible.

19.4.6 Run Samples and Acquire Data

We are now ready to run the chromatograph and collect our data from our sample injections – press go, check everything is *ok* and go home for the night. Of course, the instrument will perform perfectly and when you come in the next day, you will be ready to check the data.

Note well. The chromatographic data files we have just generated from the analytical runs are traditionally considered as the "raw data". Please ensure that you consider a wider definition of electronic records after reading this chapter.

19.4.7 Interpreting the Data Files

Ideally, if your data system has been set up correctly, you will be interpreting the data automatically and then checking that baselines and the like have been correctly positioned by the system. However, real life does not always work on autopilot and in some instances, if the chromatography has moved over the time of the run, there will be manual intervention. The data system should record if the baselines have been set by the software or manually reset by the chromatographer. Again, the interpretation is an important part of the analytical record and needs to be captured by the data system, either as a different integration file version even to an individual injection or associated with the original file from the injection. Guess what? You have more electronic records!

After the interpretation, there will be calculation of system suitability parameters and acceptance or rejection of the run, checking that the method is within calibration using the injected standards and any quality control samples are within acceptable values. This process of evaluation of the run must be captured by the data system. Yes, more e-records.

Usually, a supervisor will check the results and may reposition baselines or interpret a sample differently to the original analyst. This will also be captured in the records of the data system. And I have not even mentioned e-signatures!

19.4.8 Post-run Calculations and Reporting

After the initial results are calculated, any post-run calculations or manipulation of the results will be undertaken such as adjusting for sample weight, dilutions, *etc.* Some of these can be applied using the original method definitions or may be based upon run-specific data. In either case, these figures need to be retained for any reanalysis of the work.

Reporting can be undertaken in some laboratories using a predefined method that is applied to every batch of analytical results. Alternatively, each run requires an *ad hoc* report dependent on the specific requirements of the sample submitter. In either case, the reporting file or template will need to be considered as part of the electronic records of the work.

19.4.9 Do You Use Macros?

Some data systems allow you to record keystrokes or pre-program some functions to carry out data analysis or manipulation using macros. Macros need to be designed, tested, validated and documented as to their correct operation before they can be used but if used they also constitute part of the electronic records of the analysis and need to be associated with the work package.

19.4.10 Audit Trail Entries

By the way, I have not forgotten the audit trail records for this system; I just did not want to get the debate out of hand. The audit trail must monitor all the data files we have discussed above and needs to be archived with the electronic records. Audit trail records if archived off-line may also need to be reinstated if data are to be reprocessed – can the system do this?

The other issue to consider here is what happens in the admittedly very rare situations that you need to reinject a sample or reanalyse the whole run? The data need to be collected and stored separately; there must be no overwriting of files and the work is recorded in the audit trail.

19.4.11 Diode Array Detectors

Phew, I bet you are glad that debate is out of the way, aren't you? We have got everything covered and out of the woods, home free and we can put our feet up and relax, can't we? Er, yes and no. If you have the equipment we discussed above, then we can now be relatively relaxed about our approach to electronic records. However, do any of you use a diode array detector (DAD)? What is the impact of electronic records here?

Let us look at this in more detail. DADs can be used in a variety of detection modes:

- single wavelength
- dual wavelength
- spectral scans
- spectral libraries and compound identification

The first case is already covered in our debate above and all we may have to consider is that the file to set up the DAD is included in the collection of electronic records we archive. The second is also relatively simple, as you will usually get two files from a single injection that are linked to each wavelength monitored.

The most interesting situations, from an electronic records perspective, are the last two cases where spectral data are collected and libraries are used for identification of compounds. Consider that most data files in a CDS will be in the region of about 30–150 KB size compared with a DAD spectral scan that can be up to 5 MB. You only need a small run of samples collecting spectral data to realise the size of hard disc space you will need. This is why several CDS systems will have the option to delete the original data file, of course, you will now understand that should not happen as you will be destroying original records. The traceability of a sample from the original spectral scan to the final report must be available and audit trailed as well.

Perhaps, collection of spectral scans should be limited to the region of the chromatogram of analytical interest. This is a very practical solution to the looming electronic records problem. Put in a very blunt way, just how many void volume spectra do you want to store for posterity, especially, if you are analysing complex organic molecules?

Of more than passing intellectual interest is the situation where a laboratory uses a DAD to identify compounds by comparing the spectrum of the eluting peak with the reference sample in the spectral library. Here the spectral library becomes part of the electronic records as it is the reference point where the decision was made that peak A is compound X.

More complex is the issue where a laboratory is adding to the spectral library over time, the instance of the library at the time of the analysis is part of the whole electronic record for the analysis. Of course, I will not bother to mention the audit trail... As an aside, those chromatographers who have been working with MS detectors in some form or other are also in the same boat as the DAD users.

19.4.12 Controlled Chromatograph with Separate Data System

Up to now, we have been discussing a data system that can acquire data from an automated chromatograph and the data system itself may control the instrument. However, consider a common configuration in many laboratories. There is separate computer control of the instrument and the data system just acquires the detector signal. What are the electronic records in this instance?

This is an interesting question that will generate much debate. Here is my view. We have now *hybrid electronic* systems as the records are stored in two separate systems, usually incompatible, which means that records cannot be stored in a central location. Let us explore this situation in a little more detail. Within the data system you will have the same e-records that we have discussed above *plus* the instrument control files with the corresponding audit trail records held on the separate PC on the computer controlling the chromatograph. This separation of the e-records between two separate systems will present problems, as some users will not think to consider that the additional PC actually has e-records. The view here may be that the method file in the data system is sufficient but this argument falls flat on its face when there is no instrument control file available from the data system.

The Part 11 Scope and Application guidance[18] provides the opportunity to define the workstation controlling the instrument as incidental to the production of paper records, if a laboratory desires. However, this is only the case if there are paper records of the instrument control file before the run has started plus documenting any corrections within the run.

19.5 Define the Electronic Records for Your System

The key issue when defining your electronic records for your chromatographic data systems is to look carefully at your software application and the associated equipment. You must document the electronic records for the way that you use the system.

To help you reach your electronic Nirvana you will need to collate the answers to the following questions in your definitions document:

- Which files in your data system are needed to set up the system to acquire data and control the equipment? These will include any methods and modifications specifically for that run and the sequence of samples that include the sample identities, volumes, dilution factors and any post-run calculations.
- Are all the defined files on the same or separate computers? If the latter, how will you be able to archive them efficiently?
- Are any calibration methods and run data available to demonstrate that the chromatograph was within specification or calibration at the time of the analysis?
- Which data files were produced from each run? Are they correctly labelled and cross-referenced to the specific version of the other files used to interpret them?
- What happens with reinjected samples? Are there any overwritten files or do you have a second version of the file?
- Which files are needed to produce results from the run? Here you will need to think in more detail about how you interpret the data files, interpret the peaks (including any manual override from the automatic operation of the system), check the system suitability results, calibrate the run with the method, calculate the results in the unknowns, apply any correction factors or post-run calculations, approve the results and report them. Do not forget the files from the repeat samples or any dilutions.
- Have you considered all eventualities? Do not forget to include any special situations such as the use of DADs, mass spectrometers and spectral libraries.
- Have you included the audit trail entries for the run as well?

Check that if you can archive all these files, you can restore them as well, *i.e.* a two-way process. Remember that the archive medium you are starting with now may not be the one at the end of your record retention period. This area along with the rest of the computer hardware is purely technology driven. If you start with CD-ROM as your medium now, you will be onto DVD or magneto-optical or some medium we have not even heard of.

There you have it, conceptually simple but the difficulty is a consistent and practical implementation. Concentrate on the electronic process and what the data system and you are both doing to identify the electronic records used or created during an analysis – then document it.

CHAPTER 20

Maintaining the Validation Status During Operational Life

After operational release comes the most difficult part of computerised system validation – maintaining the validation status of the system throughout its whole operational life. The key challenge is change control.

20.1 What do the Regulators Want?

20.1.1 FDA GMP Predicate Rule Requirements

§211.68(b)[12]:

Appropriate controls shall be exercised over computer or related systems to assure that changes in master production and control records or other records are instituted only by authorized personnel.

20.1.2 EU GMP Annex 11

Clause 11[27]:

Alterations to a system or to a computer program should only be made in accordance with a defined procedure which should include provision for validating, checking, approving and implementing the change. Such an alteration should only be implemented with the agreement of the person responsible for the part of the system concerned, and the alteration should be recorded. Every significant modification should be validated.

20.1.3 PIC/S Guidance for GXP Systems

Table 28 lists some of the items that an inspector could look at during a visit to your laboratory; it is taken from the PIC/S Guidance on Computerised Systems in GXP Environments.[31]

Table 28 *Change control requirements from PIC/S guidance*

---

18 Change control and error report system (PIC/S guidance 2003[31])

18.1 The formal change control procedure should outline the necessary information and records for the following areas:

Records of details of proposed change(s) with reasoning
System status and controls impact prior to implementing change(s)
Review and change authorisation methods
Records of change reviews and sentencing (approval or rejection)
Method of indicating "change" status of documentation
Method(s) of assessing the full impact of change(s), including regression analysis and regression testing, as appropriate
Interface of change control procedure with configuration management system

18.2 The procedure should accommodate any changes that may come from enhancement of the system, *i.e.* a change to the user requirements specifications not identified at the start of the project. Or alternatively a change may be made in response to an error, deviation or problem identified during use of the system. The procedure should define the circumstances and the documentation requirements for emergency changes ('hot-fixes'). Each error and the authorised actions taken should be fully documented. The records should be either paper based or electronically filed

18.3 Computer systems seldom remain static in their development and use. For documentation and computer system control it should be recognised that there are several areas that would initiate change or a review for change. These are:

A deviation report;
An error report; or
A request for enhancement of the computer system;
Hardware and software updates

20.1.4 OECD GLP Consensus Document on Computerised Systems

Section 7c[28]:

Change control is the formal approval and documentation of any change to the computerized system during the operational life of the system. Change control is needed when a change may affect the computerised system's validation status. Change control procedures must be effective once the computerised system is operational.

The procedure should describe the method of evaluation to determine the extent of retesting necessary to maintain the validated state of the system. The change control procedure should identify the persons responsible for determining the necessity for change control and its approval.

Irrespective of the origin of the change (supplier or in-house developed system), appropriate information needs to be provided as part of the change control process. Change control procedures should ensure data integrity.

20.1.5 Regulatory Requirements Summary

Some of the key concepts from the regulations on change control are:

- Formal process and formal approval
- Scope covers both the computer system *and* the associated documentation (both written in-house and by a vendor)
- Formal evaluation of the change to understand its impact on the system and the users
- Does a change require training the users to handle the change?
- Does the change invalidate your data?

The regulator's requirements when viewed above are very clear. We just have to turn this into practice.

20.2 Change Control and Configuration Management

When I carry out audits of any operational computer system, the starting point is the changes made to the system over the period of time that the system has been running. The reason is that most computer systems change over time for the reasons listed below in this section. Changes always occur and as few systems remain in their initial configuration for long, it is essential to track all modifications to a system over time. The original purpose of many quality guidelines is being able to repeat conditions under which the work was originally done.

Look at the challenges that will be faced when dealing with maintaining the validation of a CDS or indeed any system. Some of the types of changes that will impact an operational CDS are:

- Software bugs will be found and associated fixes installed
- Application software, operating system, plus any software tools or middleware used by the CDS will be upgraded
- Network improvements: changes in hardware, cabling, routers and switches to cope with increased traffic and volume
- Hardware changes: PCs and server upgraded or increases in memory, disk storage, *etc.*
- Interface to new applications, *e.g.* spreadsheets or laboratory information management systems (LIMS)
- Expansion or contraction of the system due to work or organisation reasons
- Environmental changes: moving or renovating laboratories

The key question that needs answering from an inspector's perspective "is there demonstrable control of these changes?" In many cases there is no control of changes and therefore the system is out of control. Let us look at what is required by both change control and the associated process of configuration management.

20.2.1 Definition of Terms

There are a number of terms we need to consider here; the first two are:

- *Change control.* The systematic process by which any change to a computerised system is proposed, co-ordinated, evaluated, rejected, or approved and implemented (including tested and revalidated as necessary).
- *Configuration management.* The system for identifying the configuration of hardware, software and firmware at discrete points in time with the purpose of systematically controlling changes to the configuration and maintaining the integrity and traceability of the configuration throughout the system life cycle.

These two terms are very closely linked and some pharmaceutical organisations have condensed them to "change management" to cover all aspects of the control of a CDS or indeed any computerised system.

Note that configuration management can also be applied to software development and refers to the control of the versions of the software units and modules as the developers write them. However, for the purposes of the discussion here we will use it only in the specific context of the configuration of the operational CDS.

Going further in defining terms for configuration management we have:

- *Configuration item.* Definition of the individual components in a configuration management system. Items can include hardware (spectrometer, server, workstation), software (application, software utilities, operating system including the service packs and patches associated with all of these) and peripherals (printers, plotters). It is very important that each configuration item is carefully defined. If too detailed the change control and configuration management process will be too resource intensive and become an administrative nightmare to operate but if set too high the information generated will be useless. The granularity of the information is important.
- *Configuration baseline.* The establishment of the initial configuration of the computerised system from the individual configuration items. If a system undergoes rapid change or there are differences between the actual configuration and configuration log, it may be necessary to redefine the baseline (often called re-baselining). Furthermore, during an audit or an inspection, one will try to reconcile the configuration items in the log with the physical and logical ones on the instrument: the two should match exactly – many do not.

20.2.2 Is it a Change or Normal Operation?

Consider the following situation for an operational CDS:

- Addition of a new user or modification of an existing user's security/access control profile
- Adding a new product or project to the data system
- Adding or removing a chromatograph to or from the system

The questions to ask are any of these items normal operation or a change to the system. Consider the addition of a new user or method/project. There are utilities within the CDS to perform these operations and these should be considered as normal operation of the system. Without doubt they must be controlled and proceduralised but they are normal operations of the application that should not be under change control.

However, the addition or removal of a chromatograph is a change as it changes the overall hardware and/or software configuration. This requires a change control request, partial validation in the case of the addition of the system and an update of the configuration log.

The bottom line in many of these situations is to use common sense that occasionally can be missing in many CDS validation efforts.

20.3 Change Control Process

A typical change control procedure is typified by the following criteria:

- Responsibilities of all parties involved are defined and known
- Managed process
- Documented process

The overall process is outlined in Figure 40.

The first part of the process is a request for change. This requires some basic information such as:

- Identification of the person who requested the change
- Description of the change
- Justification for the change
- Date of the change request

The request for change may result from a variety of reasons. It may be the reporting of a bug or feature of the system software that should be resolved, the performance of the system is not adequate or there is a request for additional resources such as a printer, workstation upgrade or extra disk space.

Whatever the change requested, it needs to be documented. The way for doing this should be as simple as possible, keeping the paperwork to the minimum and to encourage all that use the system to comply with the process. An alternative to paper that could be used is e-mail with standard change request forms. Although remember that this could generate electronic records depending on how you use the system.

Second, the request needs to be analysed for its impact. There are a number of facets to consider here: the effect of the change for its impact on the laboratory and the organisation and also on the system itself. In looking at the impact of the change on the laboratory, one should consider:

- Time required to implement the change
- Cost of the change (including writing any documentation, any associated retraining caused by the change and also the time to test the change including

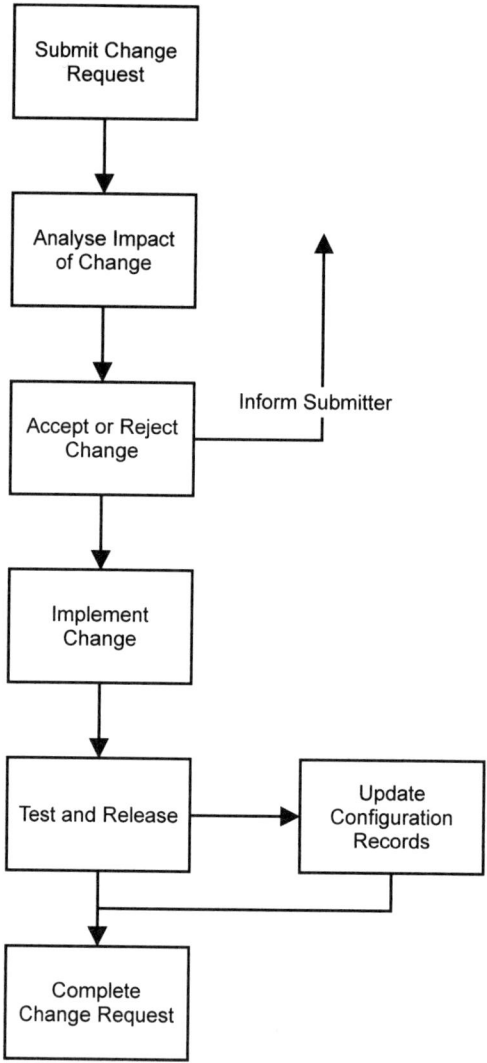

Figure 40   *Outline of a change control process*

any regression testing of the rest of the software to assess if the update has
impacted another part of the application)
- Resources required (both physical and human) to make the change
- What is the risk of making the change *versus* the risk of not making the
  change? Each change request can be classified into essential to minimise
  regulatory risk, provides business benefit or other categories
- Benefits of making the change

When looking at the impact of the change on the system, consider:

- Does the change provide a major or minor business benefit?
- Must the change be made for compliance reasons alone?
- Is the change for cosmetic reasons only?
- Is there any impact on the system?
- Are the functions already available or is an enhancement necessary?
- If the change is implemented will it cause any problems? (*e.g.* training, documentation, *etc.*)
- How much retesting and revalidation is required?
- What is the cumulative impact of incremental changes since the last full validation of the system (this is very important)?

What is the effect of the change on the organisation?

- Does the change bring a cost saving to the organisation or is more cost required?
- Will the change allow for time or cost savings?
- What impact will the change have on the documentation of the system?
- What impact will the change have on the users of the system: will there be any necessity for retraining?
- What is the impact and cost of doing nothing?

Once the impact analysis has been completed, the system owner with IT (if the system is networked) and QA can review each change. Alternatively, this can be devolved to a small validation or change control team consisting of two or so individuals authorised to consider and recommend changes. The size of the system, the business benefit and the magnitude of the change should decide the approach.

Changes should be reviewed and classified into those that bring major, minor or no business or regulatory benefits. The prioritisation of authorised changes will probably need to be balanced with the available budget and resources, as it is unlikely that all authorised changes will proceed. There will inevitably be change requests that will be rejected for a variety of reasons. Regardless of the decision by the reviewing group, it is of great importance that decisions and the rationale for making them are fed back to the requester. If the change is rejected, the submitter will be informed of the rejection and the reason for it.

However, if the request is approved, the resources are made available to implement the change. The first stage is to formulate a plan to implement the change. This will incorporate any relevant aspects of the impact assessment and any technical issues such as the extent of retesting and revalidation of the system update of documentation and retraining of users, *etc.*

The change is then made, reviewed and the system released for use. Is there something that has been forgotten here? Will a user make changes to a live and operational system? You should consider a test environment, for some systems a spare PC that you can evaluate the change and then complete the validation, on the operational system. Before implementing the change make sure the operational system is completely backup before you start, so that you have a fall back position in case anything goes wrong.

20.3.1 Discussion of Some Typical System Changes

Figure 41 is a stylised view of a networked system with a workstation controlling a chromatograph where data are stored on a network server. Both the server and workstation consist of PC hardware, the operating system and the CDS application software. Through this figure, we can illustrate and discuss any changes to the overall system to illustrate their impact.

Consider the following possible changes to the system and the impact that each could have:

- Changing the workstation? In this situation, the whole application software needs to be reinstalled and therefore extensive revalidation of the system needs to occur. The chromatograph itself will not be impacted so the instrument will not have to be requalified. However, the control of the instrument by the new installation will need to be demonstrated.
- Installing a service pack or patch for the CDS application? The release notes supplied by the vendor (they have supplied one?) will outline the nature of the extent for change when the patch is installed. Typically, you will test the functionality of the patch works for your application. In addition, you can also undertake some regression testing of the main functions of the application such as if you re-interpret a data file will you get the same results?
- Installing a new version of the CDS application software? When this happens it will typically be a complete revalidation of the system. Check that a service engineer has not also installed a new firmware version at the same time.

The impact that each potential change could have on the validation status must be assessed. For example, the hardware change that included a processor upgrade has a relatively small impact compared with the upgrade of a service pack for the operating system or a new version of the system software.

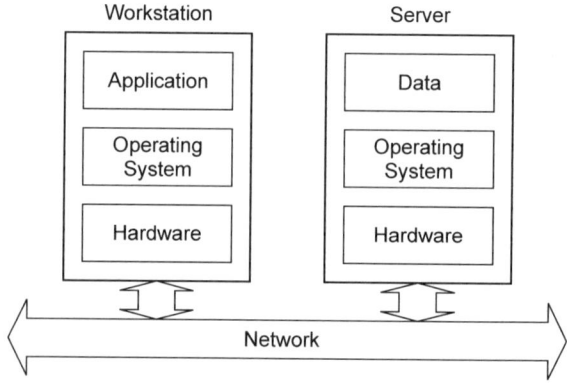

Figure 41 *Outline configuration of a networked CDS*

20.3.2 Emergency Changes

No all changes can be planned. There may be time when your software fails due to a software bug or virus, for example. Then the change control process needs to have a section dealing with how to handle emergency changes. Typically, this will allow a few authorised people the ability to make the changes without filling in the change control form and get the system running again. Then the formal documentation is completed and approved retrospectively.

20.4 Configuration Management

Configuration management, as we defined above, is a set of procedures to ensure adequate identification, control, visibility and security of *any* changes made to:

- Chromatograph instrument firmware
- Software including any patches and macros
- Computer hardware including the A/D units, data servers, workstations and server
- Peripherals, *e.g.* printers and plotters
- System documentation (specifications, procedures, *etc.*)

Furthermore, all modifications should be authorised before a change is made and the personnel making the changes should also be authorised to do so by management *via* the change control process as outlined above. Therefore, configuration management and change control are very closely linked.

20.4.1 Defining the Detail of Configuration Items

How much far do we need to go when we define the detail associated with configuration items? Let us look at what we could do for a portable PC. Here are some options that we can assess their usefulness:

- Toshiba Tecra M1
- Toshiba Tecra M1
  1.6 MHz Centrino processor
  512 Mb RAM
  80 Gb hard drive
  CD-ROM RW
  Operating System Windows XP Pro, Service Pack 1
  Serial Number 553217886 TBY
- Toshiba Tecra M1
  1.6 MHz Centrino processor
  512 Mb RAM
  Teac DZ 990T 80 Gb hard drive
  Teac DW 224E CD-ROM RW
  Operating System Windows XP Pro, Service Pack 1

Serial Number 553217886 TBY
BIOS version 1.20
Display 1068×764 resolution

The aim of these three options is to discuss at how far do you need to go? Option one is very simplistic and you cannot go into much granularity, if anything is changed, *e.g.* memory increased you cannot tell, as there is no detailed baseline configuration information to compare with. The PC could be swapped with an equivalent model and you will not know.

Option two gives more information that is relatively easy to collect and maintain. The PC is uniquely identified *via* its serial number so that you know the system is the right one. The information here though, does not tell you about any security patches that have been applied to the operating system.

Option three is more detailed compared with the other two. However, you will have to spend more time collecting and maintaining the configuration, which may be unrealistic. Therefore, configuration management information based around option two is adequate for defining the configuration item as a general rule.

There may be exceptions, where specific boards or hardware is added, to control a chromatograph these of course will need to be added to the configuration item list.

20.4.2 Defining the System Baseline Configuration

This is the process of compiling the list comprising all the configuration items components of the system that will include:

- All the release numbers, and serial numbers (where appropriate) of the application software program(s)
- The software tools (*e.g.* database), and the operating system
- The components comprising the hardware should be used such as disks, memory, type of central processing unit, add-in boards for the application or communications
- Data servers and/or A/D converter units
- Chromatographs
- Any peripherals
- System documentation should also be included in the configuration management log

The baseline configuration should be established at the installation of a new system. This has a number of advantages. First, all testing and training takes place in a controlled environment and second the procedures and principles of configuration management are known and understood, modified if necessary, before the system is rolled out for operational use. The information for the baseline configuration will come from the purchase order, and this will be checked off at the installation.

20.4.3 Linking Configuration Management with Change Control

When a change control request is approved and implemented, the change may replace or change a configuration item. Then the new configuration and the date from which it is effective is noted in the log. When new versions of the software are available and installed master copies of the old version and the relevant documentation should be archived, as they should be considered equivalent to raw data.

20.5 Operational Procedures and Records

To document the basic operations of the computer system a number of procedures and logbooks are required. The term logbook is used flexibly in this context. The actual physical form that the information takes is not the issue, rather the information that is required to demonstrate that the procedure actually occurred. The physical form of the log can be a bound notebook, a pro forma sheet, a database or anything else that records the information needed, as long as security and integrity of the records (paper or electronic) are maintained.

The following areas need to have procedures and associated logbooks or records:

- Problem reporting and resolution
- Software errors and maintenance
- Backup and recovery of data
- Archive and restore of data
- Maintenance of hardware
- Disaster recovery (business continuity planning)

20.5.1 Problem Recording and Recovery

During the operation of a computer system, boot up, backup or other system functions, it is be inevitable that errors occur. It is essential that these errors are recorded and the solution to resolve it also documented. Over time, this can provide a useful historical record to the operation of the computer system and the location of any problem areas in the basic operation.

Areas where this may be the case may be in peripherals where a print queue has stalled. This is relatively minor, however, there may be cases where the application fails due to a previously undetected error. In the latter case, there is a need to for link the error resolution to the change control system.

20.5.2 Software Error Logging and Resolution

As it is impossible to completely test all of the pathways through CDS software or any software,[39] it is inevitable that software errors will occur during the operation of the system. These must be recorded and tracked until there is a resolution. The key elements of this process are to record the error, notify the support group (in-house or vendor), classify the problem and identify a way to resolve it.

Not all reported problems of a CDS will be resolved. They might be minor and have no fundamental effect on the operation of the system and may not even be fixed. Alternatively, a work around may be required which should be documented, and sometimes retraining may be necessary.

In response to the software problem, the CDS vendor may issue a patch or service pack to resolve the issue. In this instance, the change control procedure will be triggered and an impact analysis carried out.

20.5.3 Maintenance Records

All GXP regulations require that any equipment used is properly maintained and in some instances calibrated. Computers are no exception to this. Therefore, records of the maintenance of the CDS need to be set up and updated in line with the work carried out on it. The main emphasis of the maintenance records is towards the physical components of a system – hardware, networking and peripherals. The software maintenance is covered under the error logging system described in the previous section and the contract with the vendor described in Chapter 11.

If the hardware has a preventative maintenance contract, the service records after each call should be placed in a file to create a historical record. Also any additional problems that occur which require maintenance will be recorded in the system log and there will need to be cross-references to the appropriate record.

Many smaller CDS have few if any preventative maintenance requirements, but this does not absolve the laboratory from keeping records of the maintenance of the system. If a fault occurs that requires a service engineer to visit, then this must be recorded as well.

On sites where maintenance of personal computers is maintained centrally for reasons of cost or convenience, maintenance records may be held centrally. The remit of the central maintenance group may cover all areas of a site or organisation including regulated or accredited as well as non-accredited groups. It is important for the central maintenance group to maintain records sufficient to demonstrate to an inspector of the work they undertake. As defined in EU GMP Annex 11,[27] the third party undertaking this work should have a service agreement and also have the curriculum vitae of its service personnel available and up to date.

20.5.4 Disaster Recovery or Business Continuity Plan

Good computing practices require that a documented *and* tested disaster recovery plan or business continuity plan must be available for all major computerised systems. It rarely is. Failure to have a disaster recovery plan places the data and information stored by major systems at risk, the ultimate losers being the workers in the laboratory and the organisation.

Disaster recovery is usually forgotten, or not considered, as "it will never happen to me". The recovery plan should have several shades of disaster documented.

- Consider the loss of a disk drive. How will data be restored from tape or backup store and then updated with data not on backup.
- The seriousness of the disaster can extend to the complete loss of the computer room or laboratory through fire or natural disaster. The problem with any major computerised system such as a chromatography data system is that users cannot use alternative means of calculation such as chart recorders, pencils and rulers as a disaster recovery plan. It typically needs to involve a limited recovery using smaller self-contained CDS such as a workstation running the same validated software. The nature of the contingency planning will balance the criticality of the system, the business process it supports and probability of a disaster occurring, the potential loss *versus* the cost of the proposed solution. Consider earthquake as a possible disaster scenario: in the UK this would not be given any serious consideration, but in California it is a probability and the debate looks at when and severity. In the latter case, earthquake planning is extensive and off-site storage is typically in a geologically stable location often a long distance from the location.
- The biggest issue facing most organisations today is not a fire in a computer/data centre, but the impact of malicious software such as a virus which in a worst case may require a complete rebuild of a server and any workstations. Therefore, business continuity is as much about good working practices to avoid the infection in the first place as well as the recovery plan itself.

Once the plans have been formulated, they should be tested and documented to see if they work. Failure to test the recovery plan will give a false sense of security.

The plan needs to be reviewed on a regular basis to assess if it is still current and applicable. Remember, we are dealing with information technology here and systems never remain static; technology refresh cycles occur typically on a 2–4 year period and the plan must be consistent with the technology used and updated to reflect this.

CHAPTER 21

Periodic Review of the CDS

To ensure that the system remains in a validated state and under control, independent audits of the system should be conducted on a regular basis depending on the criticality of the data system. Observations by the audit team will be noted and the system owner will be responsible for correcting the major findings.

21.1 What do the Regulators Want?

21.1.1 PIC/S Guidance

Guidance to the Inspectorate in Section 23 is an important indicator of what questions will be asked during an inspection. The following sections of the PIC/S guidance are pertinent[31];

23.9 The firm's validation approach should follow a life-cycle methodology, with management controls and documentation as outlined in this guidance, which contains consensus best practice guidelines

23.10 Inspectors should review the firm's Validation Summary Report52, (VSR) for the selected system and refer as necessary to the System Acceptance Test Specification and lower level documents. They should look for evidence that the qualification testing has been linked with the relevant specification's acceptance criteria, viz: PQ versus URS ...

23.11 Inspectors should look for the traceability of actions, tests and the resolution of errors and deviations in selected documents. If the firm has not got proper change and version controls over its system life-cycle and validation documents, then the validation status is suspect.

23.12 Inspectors should consider all parts of PIC/S GMP Annex 11 for relevance to particular validation projects and in particular, the 'Principle' and items 1, 2, 3, 4, 5 and 7.

23.13 The lack of a written detailed description of each system, (kept up-to-date with controls over changes), its functions, security and interactions (A11.4); a lack of evidence for the quality assurance of the software development process (A11.5), coupled with a lack of adequate validation evidence to support the use of GMP related automated systems may very well be either a critical or a major deficiency. The ranking will depend on the inspector's risk assessment judgement for particular cases. (NB. Since 1983, the GMPs have called for validated electronic data-processing systems and since 1992 for the validation of all GMP related computer systems).

23.14 If satisfied with the validation evidence, inspectors should then study the system when it is being used and calling for printouts of reports from the system and archives as relevant. All points in Annex 11 (6, 8-19) may be relevant to this part of the assessment. Look for correlation with validation work, evidence of change control, configuration management, accuracy and reliability. Security, access controls and data integrity will be relevant to many of the systems particularly EDP (i.e.: Electronic Data Processing) systems.

21.1.2 ICH Q7A GMP for Active Pharmaceutical Ingredients

Section 12.6 Discusses periodic review of validated systems[16] in section 12.6:

Systems and processes should be periodically evaluated to verify that they are still operating in a valid manner. Where no significant changes have been made to the system or process, and a quality review confirms that the system or process is consistently producing material meeting its specifications, there is normally no need for revalidation.

21.1.3 Regulatory Requirements Summary

Regulations require that the CDS is controlled and it is important that internal audits or periodic reviews, whether mandated by regulation or corporate requirements, ensure that the system maintains its validation status.

21.2 Rationale for a Periodic Review

A periodic review is another name for an internal audit of the CDS to assess if the system is still under control and remains validated. If there are issues corrective actions are proposed that when implemented will bring the system back under control. It is cheaper and more efficient if any issues with an operational system are found and corrected by a company than the regulators find them.

21.2.1 Who Performs the Review?

There are two options for performing the periodic review:

- System owner or somebody delegated by the owner
- Independent qualified auditor either from another functional area or QA individual

The only option for an objective assessment of the system's compliance status and adherence with written procedures is an independent. Review by the system owner or an individual delegated by him or her cannot be objective as they are typically involved with the system and its operating procedures and therefore cannot be neither objective nor independent; furthermore findings and proposed corrective actions could be ignored or watered down and make them ineffective.

The independent auditor arranged through the QA group (a member of QA, an individual in another functional group but acting on behalf of QA or an external individual) can be objective and does not have any vested interests. The review or audit conducted by or via QA means that any observations, finding or corrective actions will not be hidden or ignored.

However, the auditor works with the System Owner and their representatives to conduct the review of the system and associated documentation.

21.2.2 How Often Should the Review Occur?

The frequency of periodic review is a topic of much debate – balancing resources versus number corrective actions. However when a laboratory is inspected by a regulator a CDS is one of the two systems that an inspector will typically know something about (the other being a LIMS). Owing to the pervasive nature of a CDS and their critical use in the majority of regulated stages in pharmaceutical research and development and manufacturing make any CDS a large regulatory target during an inspection.

Furthermore, the system itself will be undergoing change with addition and modification of instruments, updates of the operating system and service packs for the CDS application that will need to be approved and controlled.

Therefore a minimum frequency of 12 months and at most a 24 month period between periodic reviews is recommended with the emphasis on the minimum rather than the maximum. Prudent companies may want to synchronise any periodic review with anticipated inspections either by their local or foreign inspections. Do not make this just a month before the inspection but six months before the inspection to allow corrective actions of critical or major findings to be implemented and operational before the inspection team arrives on site. The worst impression you can create during an inspection is present documentation that was approved the week before the inspection started.

If a new system will undergo much change in the period between initial validation and complete roll-out to a whole laboratory, then a periodic audit within six months of the completion is highly recommended. The rationale for this is that any non-compliant or incomplete practices can be caught early in the operational life of the system and corrected early rather than up to two years later. The issue is easily demonstrated by reference back to the cost of compliance versus the cost of non-compliance outlined in Figure 16. The organisation will find the cost of compliance in undertaking shorter periods of review to be far more cost-effective than a knee-jerk response after an adverse inspection report.

21.3 Overview of the Periodic Review Process

21.3.1 Periodic Review Objective

The main objective of a periodic review is to assure management and the system owner that the system remains validated and under control during its operational life. If this is the second or later periodic review then the agreed corrective actions from previous reviews must also be assessed to see if they have been implemented and are effective.

21.3.2 Planning the Audit

The review should be planned and dates agreed in advance with all involved with the process. Typically there will be a plan for the review stating what will be reviewed and checked that should be prepared by the reviewer and approved by both QA and the system owner.

The plan will outline the scope of the audit and the areas to be reviewed. This allows the system owner and their representatives to prepare for the review and retrieve any documentation from off-site storage. The agreed and approved plan also eliminates any misunderstandings about the aims and scope of the review.

21.3.3 Schedule for the Review

A periodic review would follow the following outline timetable:

- Opening meeting to introduce all personnel involved with the audit and discuss the timetable and make any last minute adjustments as a result of personnel availability
- Overview of the system to acquaint the review team with the system
- Review of the system according to the plan
- Reviewer's closed meeting to discuss the audit and plan for the closing meeting
- Closing meeting where interim finding are discussed with the system owner and their representatives

Note that there is no timetable written here, that depends on the size of the system and the extent of the review; it can last one or more days.

21.3.4 Scope of the Review

The following items indicate some of the items that should be covered in a periodic review or internal audit:

- *Security and Access Control.*
  The review aims to see if user security is accurate and reflects the current user base. For example have ex-user accounts been disabled and in a timely manner or are they still active? Does a selection of the current users on the system have user types that match their records from the SOP for account management? In particular pick users that have recently changed positions and roles within the user base.
- *Change Control and Configuration Management*
  Check that this critical SOP is followed by selecting a number of change requests and follow them through the procedure to see if the request and the procedure match. Link this part of the review with the associated configuration management records: do the changes where configuration items have been changed reflect the change request. Also look in the laboratory for a configuration item and check to see that the configuration records are accurate. Look at the overall number of emergency changes: if there are a relatively large number $>5\%$ in a young system and $>1\%$ in a mature system then there is an issue: either the system has a number of problems and needs to be investigated

further and/or the change control system is being abused. Check also the assessment of how much testing that a change requires and if this is appropriate with the scale of the change.

- *Training Records*
Current user training records with any incremental changes since the last review or go live. Are records current and up to date?
- *System Documentation*
User documentation: the URS needs to be updated to reflect new versions of the system and the current URS must be linked to the current version of the operational software
System documentation: old versions are archived and only the current versions are available for use
Procedures: are these up to date and do they reflect current working practices? Where electronic document management systems are used to manage and distribute SOPs electronically look to see if there are any uncontrolled and outdated copies in laboratory and office areas.
- *Operational Logs*
Maintenance logs of the system should be checked to see if the follow the applicable SOPs. Problem logs either as a hardcopy or via the help desk records should be reviewed over a time period to see if there are any recurring problems and what has been done to correct them. Occasionally a problem resolution can invoke a change request – specifically track this issues from problem recording into the change control process and assess if there is any breakdown of communication between the two processes.
- *Backup and Recovery*
Review the backup logs to see what problems have occurred over a given period, for example if a backup has failed did the resultant corrective action follow the SOP and is it documented adequately. The reviewer may also want to look at the training records of the IT personnel involved here and also media management. Request the log for recovery requests to see how often there are requests for recovery from tape. This may indicate data loss issues that will need to be investigated further.
- *Changes in Operational Environment*
Rapid organisational change (re-organisations, mergers and acquisitions) is inevitable in today's laboratories. Therefore assess during the audit what has changed since the last time the system was reviewed or since go-live. Is the system description current and are the users still the same laboratories or even the same organisation? Is the system owner still the same individual? If there are any changes in operation are the electronic records generated by the system still the same or have these changed and is this reflected in the documentation?

This is a small selection of the options that could be inspected during an periodic audit, others may be selected by the operations that the system carries out or the type of regulatory inspection anticipated.

21.3.5 Reporting the Review and Follow-up

At the end of the review, the auditor will prepare a report of what was reviewed and if there any findings that require corrective actions. The report is important as it is a formal record of the review and places the responsibility onto the system owner to respond. This report must be treated as in internal QA report and not shown to regulatory inspectors.

The report should list what was reviewed in depth, areas where there was a brief review and where there was no work done. This allows any future audits to see what was done and plan accordingly.

The findings should be classified as to their severity, a typical grading scheme may be:

- Critical: item that requires rapid resolution as it may result in a regulatory citation or impact the quality or integrity of data generated by the system
- Major: Does not impact the quality or integrity of data but is a serious non-compliance and should be corrected
- Minor: an issue that if left may become major or critical or is a minor non-compliance that does not impact the system or its operation directly

An action plan with the non-compliances should be another outcome of the report with suggested resolution(s) written by the auditor. The System Owner should respond to these with actual corrective actions and the timescale in which they will be completed.

Some organisations may require the System Owner to provide documented evidence that the corrective action has been completed. For example if a procedure is to be updated a copy of the new version is provided to the reviewer. Alternatively, this will be an item on the plan for the next periodic review.

21.3.6 Confidentiality of the Periodic Review Report

Periodic reviews should be treated as internal quality assurance reports and there should not normally be shown to a regulatory inspector. The FDA have produced a Compliance Policy Guide to this effect (CPG 7151.02)

CHAPTER 22

Records Retention

Retention of records is important, as this is a requirement under existing GXP predicate rules written when paper was king. The lifetime of the electronic records generated by a CDS usually outlives the system that generated them. Records retention is the biggest problem facing any computerised system.

22.1 What do the Regulators Want?

22.1.1 GLP Regulations: 21 CFR 58

§58.195(b)[14]: *Except as provided in paragraph (c) of this section, documentation records, raw data and specimens pertaining to a non-clinical laboratory study and required to be made by this part shall be retained in the archive(s) for whichever of the following periods is the shortest:*

(1) A period of at least 2 years following the date on which an application for a research or marketing permit, in support of which the results of the non-clinical laboratory study were submitted, is approved by the FDA. This requirement does not apply to studies supporting investigational new drug (IND) applications for investigational device exemptions (IDE) or applications for investigational device exemptions (IDE), records of which shall be governed by the provisions of paragraph (b)(2) of this section.

(2) A period of at least 5 years following the date of which the results of the non-clinical laboratory study are submitted to the FDA in support of an application for a research or marketing permit.

(3) In other situations (e.g. where the non-clinical laboratory study does not result in the submission of the study in support of an application for a research or marketing permit), a period of at least 2 years following the date on which the study is completed, terminated, or discontinued.

22.1.2 GMP Regulations: 21 CFR 211

§211.180 (a)[12]:

Any production, control, or distribution record that is required to be maintained in compliance with this part and is specifically associated with a batch of a drug product shall be retained for at least 1 year after the expiration date of the batch or, in the case of certain OTC drug products lacking expiration dating because they meet the criteria for exemption under §211.137, 3 years after distribution of the batch.

The key sections of this part of the GMP regulations[12] for CDS laboratory data are:

Laboratory records shall include complete data derived for all tests necessary to assure compliance with established specifications and standard, including examinations and assays as follows:

A complete record of all data secured in the course of each test, including all graphs, charts and spectra from laboratory instrumentation, properly identified to show the specific component, drug product container, in process material or drug product and lot tested.

A record of all calculations performed in connection with the test, including units of measure, conversion factors and equivalency factors.

A statement of the results of tests and how the results compare with established standards of identity, strength, quality and purity for the component, drug product container, closure, in-process material or drug product tested.

22.1.3 GMP Regulations: 21 CFR 820

§820.180 (b)[15]:

Record retention period. All records required by this part shall be retained for a period of time equivalent to the design and expected life of the device, but in no case less than 2 years from the date of release for commercial distribution by the manufacturer.

22.1.4 21 CFR 11 Requirements

Subpart B: Electronic records: §11.10(c)[17]:

Protection of records to enable their accurate and ready retrieval throughout the records retention period.

22.1.5 Part 11 – Scope and Application Guidance

The following is quoted from the August 2003 final guidance for industry with respect to the enforcement discretion for electronic records now being granted under 21 CFR 11.[18]

The Agency intends to exercise enforcement discretion with regard to the part 11 requirements for the protection of records to enable their accurate and ready retrieval throughout the records retention period (§ 11.10(c) and any corresponding requirement in §11.30). Persons must still comply with all applicable predicate rule requirements for record retention and availability (e.g. §§211.180(c),(d), 108.25(g), and 108.35(h)).

We suggest that your decision on how to maintain records be based on predicate rule requirements and that you base your decision on a justified and documented risk assessment and a determination of the value of the records over time.

FDA does not intend to object if you decide to archive required records in electronic format to non-electronic media such as microfilm, microfiche, and paper, or to a standard electronic file format (examples of such formats include, but are not limited to, PDF, XML, or SGML). Persons must still comply with all predicate rule requirements, and the records themselves and any copies of the required records should preserve their content and meaning. As long as predicate rule requirements are fully satisfied and the content and meaning of the records are preserved and archived, you can delete the electronic version of the records. In addition, paper and electronic record and signature components can co-exist (i.e. a hybrid situation) as long as predicate rule requirements are met and the content and meaning of those records are preserved.

22.1.6 FDA Inspection of Pharmaceutical Quality Control Laboratories

This FDA guide was issued in July 1993; in Section 13 of the document[68] covering laboratory records are the following:

Expect to see written justification for the deletion of all files.

22.1.7 OECD GLP Consensus Document

Section 9 – Archives[28]:

The GLP Principles for archiving data must be applied consistently to all data types. It is therefore important that electronic data are stored with the same levels of access control, indexing and expedient retrieval as other types of data. Where electronic data from more than one study are stored on a single storage medium (e.g. disk or tape), a detailed index will be required.

It may be necessary to provide facilities with specific environmental controls appropriate to ensure the integrity of the stored electronic data. If this necessitates additional archive facilities then management should ensure that the personnel responsible for managing the archives are identified and that access is limited to authorised personnel. It will also be necessary to implement procedures to ensure that the long-term integrity of data stored electronically is not compromised.

Where problems with long-term access to data are envisaged or when computerised systems have to be retired, procedures for ensuring that continued readability of the data should be established. This may, for example, include producing hard copy printouts or transferring the data to another system.

No electronically stored data should be destroyed without management authorization and relevant documentation. Other data held in support of computerised systems, such as source code and development, validation, operation, maintenance and monitoring records, should be held for at least as long as study records associated with these systems.

22.1.8 Regulatory Requirements Summary

There are a number of issues to look at with records retention:

- Complete chromatographic data, as required by GMP, need to be retained as evidence of the work carried out for the duration of the records retention period. This should include all changes and manipulation of the data.
- The records retention period varies with GXP discipline; the minimum is batch expiry plus 1 year for manufacturing data or 2 years after an NDA approval. However, there are potential complications in product liability legislation limits (11 and 20 years in Europe and the US, respectively), the need to support additional licence applications with existing data and the electronic Common Technical Document (eCTD) mentioning that product licence data may need to be kept for up to 50 years.[69]

- There is the option, based on a documented risk assessment, to convert from the original electronic format to either electronic format or paper provided that the content and meaning of the records is preserved.
- Archived data (either paper or electronic) must be secure and organised for easy retrieval. Note that this is not instant replay. If data are archived off-line replay cannot be instant.
- When records are deleted there must be a formal documented process authorised by management.

22.2 Impact of the Lack of Universal CDS Data Standards

During the 1990s a standard data file format for CDS files was developed; data generated on one CDS could be read and interpreted by a different application. The underlying file format is based on the network Common Data Format (net-CDF) which was used as a means of transferring astronomical data between observatories; details of the file format are available in Liscouski's book.[70]

However, the emphasis of the standards work was on the chromatographic data file only and not on the surrounding metadata (instrument control file, method, sequence file, report, *etc.*). Therefore, although the data file can be transferred and read by another data system, it is unlikely that any metadata can be read by another CDS, as there is no common standard for these files. The only way this information can be transferred between different data systems is manually. With the large volumes of data generated by a CDS over time means that manual transfer is not a practical option. However, it is not a good option for records retention, as the file format net-CDF does not allow replay of the data for the reasons above.

Until we have universal data formats that allow true interoperability between different CDS applications from different vendors, you are effectively locked into a single vendor who has the responsibility of either:

- maintaining stable file formats, or
- providing fully working and documented conversion routines to enable data migration between one version of file and another

This is an important reason for ensuring that the system selection process is got right first time.

22.3 Options for Electronic Records Retention and Archive

22.3.1 Archive is Different from Backup

It is important to differentiate between backup and recovery and archive and restore (or retrieval); for the purposes of this book the terms are defined as:

- "Backup and recovery" is used for short to medium term data storage where there is a need for immediate recovery of data (from files to disks) in case of file loss, corruption or system failure. This is a pre-requisite for disaster recovery or business continuity planning.

- "Archive and restore" is concerned with the long-term storage of selected data. In this context, the use of "store" is also used to denote archived electronic records.

Backup is a regular process usually occurring daily and performed by the computing or IT (Information Technology) department. It is concerned with the whole of the system or the data disk. In contrast, archive and restore occurs infrequently and is driven by the users (the archiving process may be carried out by the IT department, but it is user specified). The data to be archived are carefully selected and will be by work package rather than the whole or part of a disk.

22.3.2 Organising CDS Electronic Records to Archive

You will need to organise the archive of your CDS data around specific packages that will depend on the type of work you do. Some typical examples could be the complete electronic records from the following types of work packages:

- Specific analytical method
- Batch records of the same material
- Stability study
- Pharmacokinetic study or protocol
- Clinical study or protocol

These are only suggestions, as most functions within organisations tend to work slightly different and you may find other approaches that will fit your specific needs better.

Organisation of archive of data concerns specific rather than general work packages because once you have archived the data you may have to get this back at some time. Spend time in designing a simple way of defining your requirements and get it right first time. Alternatively, you can hope you have retired or the company has merged and it is someone else's problem. The problem is that you could wait for a long time before it emerges that you cannot easily restore data, as they are stored in several places or in the worst case, the system does not work at all.

When defining the electronic records to archive be sure to include the pertinent audit trail records as well to comply with the specific requirements of 21 CFR 11.

22.3.3 Options for Electronic Archive

There are two main schools of thought about archive and restore of CDS electronic records: on-line storage and archive off-line to alternative media.

One approach for archive and restore preferred by some IT professionals is that we should archive data on-line rather than off-line. As hardware is more resilient and fault tolerant, therefore appropriate servers and disk space can be purchased relatively cheap. As the electronic records that we generate with a CDS grow in number and volume, we must keep pace by purchasing extra disk storage. As we approach the disk capacity of the server, simply purchase a new server and transfer the records. However, there can be problems with this approach as there can be uncertainty with defining the electronic records and associated metadata to transfer to the archive.

System performance may be an issue. Some points to consider as the number and volume of electronic records and data files increase are:

- Can the file management or database system cope with the increasing size of the electronic records over the lifetime of the application use?
- Was this approach tested and supported by the vendor?
- Are the security and integrity of the records maintained? For instance, is it easy to change records that are approved or final?
- What happens if there is a disaster? Can you recover all of the records? Has this been tested?

My view is that you should keep records on-line for a certain amount of time and then archive them off-line. The length of time before the electronic records are archived will depend on the nature of the work in which you are involved. Some clinical/bioanalytical and product stability studies may last 2–5 years and thus after the report has been prepared and authorised may be a suitable time to consider archiving the data. However, if the data must be accessed later then the decision may need to be delayed.

My reason for suggesting the records be archived off-line is partly due to protection of records and partly in case of disaster. The records are the organisation's intellectual property and will be the means of meeting regulatory requirements during an inspection. Therefore, they must be protected. Unless there are stringent controls on access to the data there may be a risk of corruption of the records. Furthermore, as the size of stored records grows, the performance may suffer if records are kept on-line.

22.3.4 Can I Read the Records?

One Part 11 requirement is for replay of the data – providing the data are in a format that they can be replayed. Note that this is *not* instant replay because with an effective off-line archive process, records will be stored in one or more remote locations and will have to be restored back to the appropriate computerised system before replay. The key issue here is to ensure that whatever archive process you use works from the computer system as well as allowing you to restore and read all the electronic records whenever you wish.

There are a number of questions concerning changes to your data system that will affect the ability to replay data. Look at any differences between when you originally acquired the electronic records and when you want to read, reprocess or replay and consider:

- Are any file formats the same?
- If your electronic records are stored in a database, is the structure the same?
- Have the integration algorithms changed for in the chromatography data system between versions?
- Has the operating system changed and can this make an impact?
- Has the application software changed?
- Has the hardware platform changed?
- Has the archive medium changed?
- Has the archive software changed?

As you can see, there are a number of changes that can have an impact on the ability to develop an effective archive. Changes in any one of the above could limit or destroy the ability to restore data from an off-line archive. The people who vote for an on-line archive will still have the same problems with many of these questions, except that they have an immediate impact (on-line) rather than a delayed one (off-line).

This area is one of the most technically challenging problems with which we are presented for compliance with 21 CFR 11. We will look at two in the next two sections, those of file format and archive media followed by the issue of data migration.

22.3.5 Impact of a Changed CDS File Format

What impact does a file format have on archive and restore? There are a number of issues to consider: however, to help you achieve a better understanding of the problems, think back when you were using a word processor.

If you have used the word processing application Word for a long time, you will remember the problems with migrating from Word 95 to Word 97. This illustrates what we will face with our electronic records but over a longer period of time. The document file formats were not the same and the migration route originally used conversion to rich text file format. The problems did not end there as the conversion was not perfect and a number of features did not work. For example, a table of contents transferred acceptably, but all the page numbers were migrated as "1". In this case, all you had to do was delete the table of contents and reinsert it, but imagine the audit trail entries for regulated data. Transpose this to your regulatory records: how would you feel if all your chromatographic peak areas came out as "1"?

Changing file format can have a major impact on the ability of a program to replay data – this can have a direct impact on long running work such as a stability study or a clinical trial. Therefore, one of the consequences from a business as well as a 21 CFR 11 perspective will be that you will tend to be locked into a specific vendor. You want to take a look at the method of conversion to ensure that the content and meaning of the records are preserved after the conversion.

22.3.6 Selection of Off-Line Archive Media

There are many problems with selecting the media used for off-line archiving; there are a number of options:

- CD-ROM has a de facto standard based on the Sony–Phillips co-development of the technology with a reasonable capacity of 650 MB unless you have large data files to archive, *e.g.* using DAD detector spectra.
- Magneto-optical drives have larger capacity and are stable against magnetic fields.
- DVD disks have larger capacity (up to 17 GB) but there are many standards and capacities and it is not known which one will succeed.

This is not very encouraging and currently, the safest approach is to use CD-ROM, as this is the best standard. However, it is a relatively mature and well-understood technology but with relatively limited file capacity which may be an issue in certain circumstances. Nevertheless, unless you are creating large file sizes, then a CD-ROM disk capacity is usually adequate for archiving most CDS archiving. However, whatever archive medium you start with, it will typically be replaced before the end of the record retention period.

22.4 Changing CDS – What are the Archive Options?

There may be an occasion when you may need to consider alternative approaches to records retention as outlined in the FDA Guidance on Part 11 Scope and Application.[18] Here is the option to transfer electronic records to paper (and other media) providing the content and meaning of the records are preserved.

If faced with changing from one CDS data system to another, then the options for preserving the records should be reviewed and assessed. What does really this mean in practice? This section should also be read in conjunction with the chapters on data migration (Chapter 23) and system retirement (Chapter 24) to get a full picture of the options available.

22.4.1 Overview of Some Options

Some of the options for consideration for the records retention of the replaced data acquisition system are:

- Maintain the old CDS records as paper and formally destroy the corresponding CDS electronic records.
- Maintain the old CDS records as paper and maintain the corresponding CDS records with a workstation to reprocess the data if there are any requests from laboratory customers or inspectors.
- Import the old CDS data files into the new data system *via* a conversion utility and manually enter the other metadata.

- Convert the records to PDF using Adobe Acrobat providing the conversion preserves content and meaning of the records.
- Convert to electronic format using a Scientific Data Management System (SDMS).

Other options may be possible depending on the two data systems involved and technical options available at the time of the work occurring.

22.4.2 Assessment of Option Feasibility

As part of the risk assessment process, each of the above options should be reviewed against a series of criteria such as:

- Technical feasibility: each option in the section above should be graded as easy, medium or difficult.
- Preserving record content and meaning: either Yes (preserves content and meaning) or No (does not preserve content and meaning).
- Cost of the approach: evaluated as either low, medium and high.
- Regulatory risk: graded as low, medium and high.
- Value of records over time can cover a number of issues: *e.g.* how often have archived records been accessed and where the laboratory is in the pharmaceutical value chain (R&D, active pharmaceutical ingredient or secondary production).

Each factor can be scored with a simple method such as technical feasibility options such as low = 1, medium = 2 and high = 3. The same principles can be applied to the other factors for consideration in the risk assessment. To help make a decision, the scores for each factor can be added up and one with the lowest total is the preferred option. However, it is impossible to give definitive guidance on what to do in an individual instance, as each situation has unique factors to consider.

CHAPTER 23

CDS Data Migration

Data migration will be necessary for a number of reasons even when the application software is not changed. It is essential to consider the impact of data migration in the following scenarios every time CDS software is upgraded or changed:

- Change in data processing algorithms following a software upgrade of an application
- Change to use of a different software application
- Change in computing environment such as operating system or computing platform
- Change in data file formats

23.1 What do the Regulators Want?

Data migration and system retirement usually occur at the end of the life cycle of any CDS. However, there are no directly stated regulatory requirements for formal system retirement with the sole exception of the 21 CFR 11 requirement[17] for preservation of electronic records. However, there is now enforcement discretion under the new Part 11 guidance[18] where alternative approaches are possible such as conversion to paper as discussed in Chapter 22. This chapter will look at data migration from the perspective of the theory and the practice with a case study.

23.2 Business Rationale for Data Migration

Data migration can be the worst part of computerised system validation as a system generating the data will have been operational for a number of years, the data may be shared between several departments and the original staff involved with the project no longer work in the organisation. This can be compounded where there have been reorganisations within a firm and the system boundaries are different compared with the original installation.

The data to be migrated can be part of the intellectual property of an organisation that needs to be protected and maintained. There may also be long term studies ongoing such as product stability or clinical trial sample analysis that may need to have chromatographic data especially when measuring low concentrations of analyte as this will be the time that system differences will be observed.

You will be in a situation in which you are at the end of one life cycle and at the beginning of another and some of the issues to face are these:

- What to do about your existing system and the accumulated data it has generated and processed, especially if there are electronic data files?
- How do you cut over to the new system and what is the impact on the work being undertaken by the laboratory?

23.3 Drivers for Data Migration and System Retirement

The drivers for data migration and system retirement are usually from two sources: internal and external or a combination of both. We will look at both to examine the reasons for this.

23.3.1 Internal Drivers

Here the replacement of a system may result for several reasons:

- *User input*: Does the current system(s) do the job required? The business function may either drift or change dramatically over time and the CDS may not meet the current requirements. Alternatively, an increase in functionality sophisticated and need more from the system to work effectively.
- *Several data systems reduced to one*: Over time a laboratory may acquire a number of data systems, either because of changes in purchasing policy or because of personal preference. This means that a single chromatographer would need to be trained on several different data systems to be effective within such an environment; this would be coupled with the different capabilities of each type of data system. Retirement of all but one, or indeed a move to another data system, would mean that a chromatographer could be trained for a single system.
- *Corporate policy*: This is when an organisation has made a decision to use a single CDS for either financial and/or standardisation reasons. Alternatively, the introduction of a standard or common office environment (SOE or COE) for the PC desktop could trigger the decision for a change to comply with a COE.
- *Business decision*: The contraction or closure of a department or even a site may necessitate the retirement of a system and also data migration to another department or site.
- *Computerised system validation policy*: Some organisations include a retirement and data-migration phase in their CSV policies and this is merely the execution of that policy on system replacement.

23.3.2 External Drivers

The rationale for system retirement can derive from such factors as:

- *Existing system obsolete*: This occurs when a vendor makes a system obsolete or changes the operating or hardware platform. The old system is declared obsolete and will not be supported after a specific date, and the laboratory has to move to a new system if it wishes to continue to receive effective support from its vendor.

- *Interpretation of existing regulations or guidelines*: This may result in an action at your laboratory from the regulating or certifying authority that means you must improve your data system. Alternatively, this occurs at a laboratory outside of your organisation; wishing to avoid similar action triggers the search for a replacement.
- *Introduction of new regulations or guidelines*: Here, an existing system does not comply with these new guidelines, and improvements in the system functionality are required if it is to comply with them, *e.g.* 21 CFR 11.[17]

23.4 Data Migration Options

There are several approaches to the retirement of a system that could be considered:

- *Turn off and forget*: As the name suggests this is turn off the old system and forget it, then start using the new CDS. There is a slight problem about the existing electronic records within the system.
- *Phased cutover*: Complete the existing work on the old system and undertake all new work on the new system. This may take a long time and will need two systems operating in parallel with all new work on the new CDS and on-going work with the existing CDS. This is not very practical.
- Migration of the data and retirement of the old system components.

Why bother to go through all this fuss over an old system? Let us look at some of the reasons for and against the argument for formal data migration and system retirement. Consider an overview of your existing system. Most CDS will have been operational in your laboratory for between 2 and 10 years. Either stand-alone or multi-use systems, they will have analysed many thousands, hundreds of thousands or even millions of samples. Some work, especially that assessing the stability of products may have run for a number of years. Cutover to a new system will inevitably involve studies assayed under the two systems.

23.4.1 Data Migration between Different Applications

The usual type of data migration involved transfer of data between two different CDS systems. This will be discussed in detail in a case study[71] that describes the experiences validating the data migration between two different platforms quantitative LCMS data systems will be presented. The triggering event for the migration was the decision by the vendor of the mass spectrometry equipment and application software to move to a new computing platform and declare the current one obsolete.

23.4.2 Data Migration within an Application

Usually data migration is only considered when you move from one system to another. However, now you will have also to consider the impact of data migration within the same vendor's system if a change is made that affects the ability of the new version of the CDS to integrate, manipulate and report *historical* data: do you get the same results? If you have a commercial CDS system, ensure that you read the release notes to see what has changed in the new version of the application. If there is something that impacts on the ability of the system to repeat the calculations and reports that you have performed earlier, then you need to consider data migration.

Equally, vendors must make efforts to reduce such changes and provide full information on the impact of the change. If there is a major impact such as a change in data calculations, the vendor must provide working and fully documented solutions or permit the old calculations to be used as a default.

23.4.3 Validation of within Application Data Migration

When a CDS application is upgraded and undergoes the appropriate level of revalidation, one of the PQ tests should be a test of what happens if historical data is reprocessed by the new application. Ideally, there should be no differences. However, this may not always be the case as software will always contain errors and therefore some of these may be in the integration and data processing algorithms of the system and these could have an impact on the reprocessing of data. Evaluate the reprocessed results carefully and assess if the system can be used. There may be minor differences between versions which must be documented, but the main way to decide if these have an impact is to assess if the same decision would be taken with the new data (*e.g.* sample meets specification). If this is true, you may probably want to implement the system. If the decision is not the same then you have a potential problem and you may not be able to implement the new version of the application.

23.5 Generic Data Migration and System Retirement Process

A generic seven-step process, shown in Figure 42, describes system retirement and migration of data. Each stage will be described in overview and is a summary of the work described by McDowall.[72] This will be an introductory discussion before detailed data migration within this chapter and system retirement in Chapter 24.

23.5.1 Role of the System Owner and Senior Management

The system owner and senior management will need to be involved in any data migration project as the individual is legally responsible for the system. Senior management will be involved because of the resources, cost and mitigation of regulatory risk that is involved.

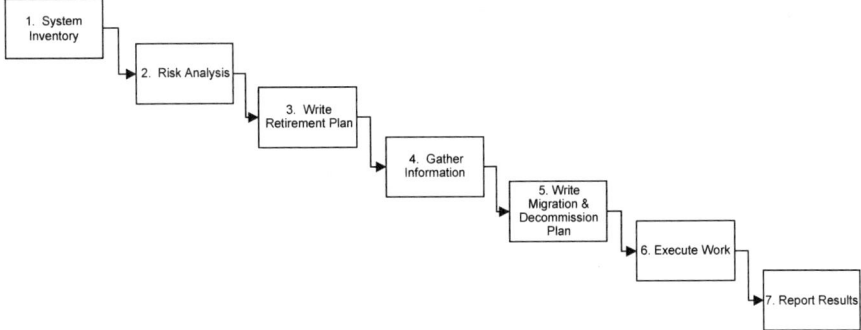

Figure 42 *Generic process for data migration and system retirement*

The system owner is responsible to the whole process, at the start the individual needs to:

- Identify the user groups (stakeholders) involved.
- Identify who owns the data on the system. This can vary from one individual to several departments, depending on the size of the overall system.
- Identify the completed work to be archived.
- Identify the completed work to be migrated.
- Identify the ongoing work to be migrated.

The system owner needs to involve all stakeholders in the system to establish a retirement and data-migration team. As mentioned above this team will report progress to senior management and typically there are decision points before the commitment of large outlay of money and resources.

23.5.2 Step 1: Inventory of the System

Identify the scope and boundaries of the system and the departments who use the system. Part of this may be the fact that the system may be spread across buildings and even networks. The latter is an issue, as it can complicate the initial work, as data spread over different networks will have to be collated to find out the data volumes and projects/studies involved.

23.5.3 Step 2: Carry out a Risk Assessment

How critical is the CDS system and the electronic records generated by it? The outcome of this assessment determines the level of regulatory risk and data criticality and is used to determine the detail required in the remainder of the process. In this process consider both the records retention period as well as the intellectual property in the CDS records – what can you afford to delete and what must you keep?

23.5.4 Step 3: Write the Retirement Plan

Using the data generated from step 1, write an overall CDS retirement plan that covers:

- Scope and boundaries of the chromatography data system(s)
- Roles and responsibilities of those involved in the data migration and system retirement
- Outline project plan
- Process of system retirement
- Process of data migration

23.5.5 Step 4: Detailed Information Gathering

In this part of the process you will need to know the details of the computer hardware including any specialised CDS devices, the software and the documentation associated with the system as well as the data. The data need to be identified in detail, for example:

- How many tapes are involved (if your long-term storage is on tape).
- How on-line data are structured and stored.
- What data relating to which samples or work packages are on a specific tape.

23.5.6 Step 5: System Decommissioning and Data Migration Plan

This document is a detailed presentation of the approach you will be undertaking on the system and describes the roles and responsibilities of people involved in the work, the systems, the data to migrate, the test scripts needed and what each test script will contain to document the process.

23.5.7 Step 6: Execute Work and Document Activities

Following the tasks described in the decommissioning plan, the data migration will start first to be followed by the system retirement. You will need to write any scripts to check and document the correctness of the data transfer; this is a critical stage in generating the confidence in the process. Once the data have been successfully migrated and or archived, then you will turn your attention to turning off the hardware and reusing it or removing it from site. This will be documented as the process continues.

23.5.8 Step 7: Write Retirement and Migration Report

This is simply a summary of the work that was undertaken with a description of any deviations from the plan and a discussion of their impact. The data migration together with any validation tests applied will be described and management will sign off the report.

23.6 Case Study of Data Migration

Turning theory into practice is described in this section; the content is based on a case study that was published previously.[71] It is based on work carried out in a bioanalytical laboratory of a contract research organisation (CRO).

23.6.1 Design of the Overall Validation Project

The project consisted of three strands of work under a single validation plan as shown in Figure 43; the strands of work were:

- Prospective validation of the new application software (Analyst) and qualification of new instruments associated with them.
- Validation of the migration of electronic records generated using MassChrom software (this name covers the data acquisition and instrument control modules called RAD or sample control as well as quantification and calibration modules called MacQuan or TurboQuan) on the Macintosh systems to the new Analyst NT environment as well as data acquisition on some Macintosh platforms with interpretation using Analyst software.
- Formal retirement of obsolete mass spectrometry (API III instruments) and obsolete Macintosh computer hardware (Quadra computers).

This chapter concentrates on the data migration strand only.

23.7 Overview of the Mass Spectrometry Systems

The mass spectrometry equipment, current software options and computing environment within Bioanalytical Services is presented below and summarised in Table 29.

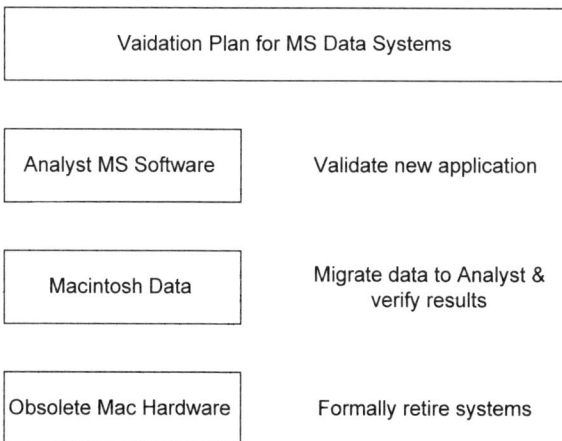

Figure 43 *Overview of the whole mass spectrometry validation, data migration and system retirement project*

Table 29 *Data processing options available in the Macintosh MS Systems*

Mass spectrometry instrumentation	Computing hardware	Operating system	Data acquisition software	MS quantification software
API III+	Mac Quadra	Mac OS	RAD 2.6	MacQuan 1.4
API III+	Mac Quadra	Mac OS	RAD 2.6	TurboQuan 1.0
API 365	Power Mac	Mac OS	Sample Control 1.3	MacQuan 1.4
API 365	Power Mac	Mac OS	Sample Control 1.4	MacQuan 1.4
API 365	Power Mac	Mac OS	Sample Control 1.4	TurboQuan 1.0
API3000	Dell PC	Windows NT	Analyst v1.0	Analyst v1.0

23.7.1 Mass Spectrometry Equipment

There are three main models of mass spectrometer currently operating in the Bioanalytical Services Department: API models III+, 365 and 3000. Of these, the API III + is obsolete as the Macintosh PC used to run the software is no longer in production. Therefore, the three systems using the API III + mass spectrometer will be formally retired and only the API 365 and 3000 models will be used thereafter.

23.7.2 Data Acquisition and Processing Software Applications

The MassChrom mass spectrometer software currently used in the department is a combination of data acquisition software (three versions of RAD and sample control) and data processing software (two versions) that operates on the Macintosh plus the Analyst software designed for the Windows NT environment. The RAD and MacQuan software running on the Macintosh Quadra will be retired under the work described here.

A mixed environment will be operated for a transition period where data are acquired by sample control running on a Macintosh but all data processing and quantification run on the Analyst. In future, after retirement of all Macintosh computers, there will be an environment that is only Analyst running on Windows.

23.7.3 Computing Environments

The current environment was Macintosh with mass spectrometry being downloaded to a server after it had been acquired. Introduction of the Analyst has started a migration to an NT operating environment that will continue after the completion of the data migration outlined here.

23.7.4 Differences between the Two CDS Systems

It is vitally important to understand the differences between the two environments before progressing further with any data migration. Covered here are the major differences between the two systems and their impact on the data migration. Essentially the problem is that we have incompatible

- Hardware
- Operating system
- Application software
- Data file formats and
- Application design philosophies

These differences will be discussed below, however the bottom line is that data file conversion is essential for the data migration to succeed.

- *Computing platform differences*: The Macintosh and Intel hardware computing platform and operating system software are essentially incompatible. An emulator is needed to run Windows software on a Macintosh, but there is no corresponding emulator for the Macintosh in a Windows environment that will run the software and be supported by the vendor.
- *Raw data file format differences*: The file formats for the chromatograms produced by the same instrument in the two environments are completely different. The Macintosh uses a different file format compared with the Analyst that uses WIFF (Waveform Interchange File Format) file format and can have either single or multiple WIFF files. For the work described here, only the use of multiple WIFF files was evaluated.
- *Meta data file format differences*: The MassChrom software (RAD or sample control) generates and maintains three files to set up and acquire data: the method, state and experiment files. The method and experiment files are used to set up and acquire mass spectrometer data and the experiment and state files used to monitor the performance of the mass spectrometer itself. In contrast, there are just two such files used within the analyst: data acquisition method (DAM) and instrument (INS) files. The mapping of the MassChrom and Analyst files is not one to one: parameters in the experiment file are split between the INS and DAM files on the Analyst application.
- *Design philosophy of the Macintosh and NT software applications*: Although the software running on the two platforms can control the same mass spectrometry instruments, their designs are very different. The MassChrom software was designed in the early 1990s for operators with mass spectrometry training. The terminology and instrument set-up within the applications are specialist for trained mass spectrometrists.

Over time the instrument has been used more widely by chromatographers and the Analyst software is a response to this as the operation of the application is simpler and uses chromatographic terms more than mass spectrometry ones. This difference in design philosophy is a complicating factor for the data migration, as terms have to be mapped between the applications, as we will describe later in this chapter.

23.8 Data Migration Strategy

The options for data migration are to assess if it is technically feasible to migrate data. The vendor of the mass spectrometry software systems (Applied Biosystems/MDS Sciex) provides conversion programs to allow a user to migrate electronic records from the Macintosh to the Analyst system. Conversion is necessary, as the file formats are completely different between the Macintosh and NT environments.

23.8.1 Vendor Supplied Data Conversion Utilities

Three API File Converter programs were supplied for the conversion of the Macintosh format data and metadata files by Applied Biosystems, the software vendor, these are:

- *File Translator*: Data file conversion program that takes Macintosh formatted data files and converts them to single or multiple Analyst format files (WIFF).
- *InstFileGenerator*: Instrument file conversion program combines Macintosh state and calibration files and generates an Analyst instrument file (INS file).
- *ExptFile Converter*: Experiment file conversion program combines a Macintosh state file and a Macintosh experiment file and generates an Analyst DAM file.

Therefore, it is technically feasible to convert the data and migrate them into the NT environment, the question now becomes "are all data converted or are files converted on an as needed basis?" The data volume involved is in the range of 100–200 GB of data.

23.8.2 Limitation of the Data Conversion Utilities

These utilities have a number of limitations that were not apparent during the early stages of this work:

- They only work on a PowerMac. Therefore, the objective of retiring all Macintosh computers cannot be realised as at least one is required to run the data conversion utilities.
- The utilities cannot convert RAD version 2.6 files. Only the chromatograms can be converted but the experiment, method and state files cannot and the data contained therein must be manually input into the Analyst. Therefore, in the case of data collected under RAD version 2.6, the requirements of 21 CFR 11 for ready replay of data cannot be met.
- A further limitation of the utilities became apparent during the data migration in that the original baselines were not transferred and new baselines redrawn with the new system.

23.8.3 Data Migration Options

There are essentially two options for the migration of the data from the MassChrom environment:

- Convert all data into the new data format now
- Convert selected data on an "as needed" basis

The second option was chosen for a number of reasons including the time and cost of conversion. However, two main issues arise from this approach. The first is that the laboratory is totally reliant on the vendor's conversion utilities and their continued maintenance of them over time and second the conversion utilities must be tested to confirm that they continue to operate as expected after every software upgrade.

23.9 Evolution of the Data Migration Design

It is important to understand that a data migration project requires a full understanding of the problem. Therefore, this section of this chapter is intended to provide a measure of the evolution of the data migration project as the understanding of the extent of the issues involved increases.

Initially, a single test script under the Analyst validation was envisioned. However, as the complexity of the MassChrom software versions was understood, a data migration and system retirement test plan was required to explain the overall strategy with five test scripts. Further information gathering revealed more complexity and the number of test scripts rose to 10.

A complicating factor was that each combination of MassChrom software had been validated on its own, comparison of data across all combinations of the software had not been undertaken as this is not normally considered as part of a normal validation study. Therefore, to ensure a comprehensive approach to the data migration, an evaluation of data acquired by all MassChrom software versions was required to ensure that no regulatory questions remained with the data migration. This approach increased the number of test scripts to 16.

Detailed design of the test scripts enabled a better way of testing to be developed and this reduced the number of test scripts down to 12, of which three were for retirement of the obsolete mass spectrometry systems.

23.9.1 Design of the Overall Data Migration and System Retirement

As there was no systematic study of results from all MassChrom software combinations, it was decided to evaluate results from all MassChrom software combinations *versus* Analyst. In addition, all future data acquisition and analysis configurations were also evaluated to give a comprehensive approach to the data migration and find out if there were any problems with the proposed approach.

Standardised Study Design. As the Analyst version 1.0 had been comprehensively validated to include some 21 CFR 11 requirements,[73] we decided that this was the standard to which all data migrations would be measured. A series of 32 sample vials were prepared containing standard and blank solutions that represented a standard curve and a series of unknown samples. This standard set of samples was injected into a mass spectrometer controlled by the Analyst software and this set of acquired data was considered the gold standard against which all data migration results were measured.

The standard sample set was then injected into different mass spectrometers controlled by the different software versions, the data analysed and then migrated into the Analyst using the vendor's utilities and reprocessed. Therefore, we have a situation where the same samples solutions have been acquired and analysed by the various MassChrom software versions and then migrated into the Analyst and reprocessed and compared against the results of the same samples acquired and processed directly by the Analyst.

In addition, historic study data acquired under MassChrom and archived on tape would be restored to the server, all electronic records then migrated to Analyst, and the results compared.

All test scripts were written, technically reviewed and then approved by the Quality Assurance Unit before execution.

23.10 Data Migration: Key Results

In this section, a selective review of the key results obtained from the data migration to illustrate the issues in a data migration project is presented. Four areas will be discussed in light of the migration issues we found, the acceptance criteria that we set, and the results that were obtained after the migration.

23.10.1 Retention Time

Retention time is a fundamental chromatographic parameter and is the time that the chromatographic column retains an analyte. In setting the acceptance criteria, the discussions centred on the conversion of time and we determined that the retention times should be within 1% of the original value, especially as the applications were both from the same software supplier. The acceptance criterion of $\pm 1\%$ was determined on the basis of a 3-min chromatographic run time and that there are likely to be differences in the peak integration algorithm that may impact the peak apex in the migrated data.

Reviewing the migrated data, it was seen that there was a large discrepancy between original and migrated results:

- 1.07 (MassChrom)
- 1.12 (Analyst)

Thus, the migration of this parameter appeared to fail against the acceptance criteria. Examining the data more closely, the data formats between the two are different: minutes and seconds (MassChrom) and digital minutes (Analyst). Therefore, we are not comparing like with like and the MassChrom values must be converted to digital minutes to make the comparison valid.

Therefore, all MassChrom retention time values must be collated, converted to seconds then divided by 60 before comparison with the corresponding Analyst values. After this conversion, the converted retention times were similar to the original results within rounding errors in the second decimal place. In retrospect, the acceptance criteria could have been set within ±0.5%.

23.10.2 Instrument Control Parameters

As mentioned earlier in this chapter, there are design differences between the two software applications and these are manifested in the instrument control parameters in both that can have no or a major impact on the data migration. This area requires a thorough knowledge of the two applications, failure to do this means that the migration will be flawed due to lack of knowledge.

For example, some parameters are the same in both applications and present no problem in the data migration project. An example of this parameter is the scan type such as multiple reaction monitoring (MRM) that is present in both applications, therefore the migration is relatively straightforward and the acceptance criteria that are set is an exact match.

However, a parameter can have different terms in the two applications but still refer to the same measurement, and this starts to complicate the migration, as the parameters must be mapped. A typical example is the Q0 voltage (MassChrom) that is equivalent to the Entrance Potential (Analyst) and illustrates the design differences between the two applications. The acceptance criteria in this instance were set to the nearest volt ignoring differences in the decimal values (*e.g.* 3.0 *versus* 3.00), the rationale was that we did not know how numbers were held in either system and that there might be rounding errors involved in the migration.

Adding further complexity to the migration is where a parameter in Analyst has to be derived from two parameters in MassChrom. Thus, the collision cell exit potential value in the Analyst can only be calculated by subtracting the potential for the Rod Offset Potential Q2 from the Inter Quad Lens 3 potential. The acceptance criteria for this were the same as the last example (the nearest volt ignoring differences in decimal values).

Again, this reiterates the need to fully understand the two applications before beginning a data migration. The acceptance criteria for all the instrument parameters monitored in the migration were documented in the appropriate test scripts that were reviewed and approved before the migration.

23.10.3 Integration Algorithms and Calculated Results

When migrating data from one application to another there are a number of results that can be compared. In the example of mass spectrometry these include:

- Analyte peak heights or areas
- Drug: internal standard ratios
- Calibration curve parameters
- Calculated results from unknown samples
- Back calculated standards

As the integration algorithms were different between the two applications, an early decision in the migration was to avoid using the peak area calculations as a comparator between the two systems as noted by McDowall[72]:

> *What we need to consider here is, when the data files are in the new data system are similar results ... obtained? Expect to see some differences between the two systems. The main issue is whether it matters from a scientific perspective ... For instance, if the final calculated result means that a sample that was previously acceptable is now out of specification, the impact of this needs to be assessed ...*

This situation was confirmed from the first set of converted data shown in Table 30.

Note that the data at first glance are very comparable, however on closer inspection the Analyst data were consistently higher. Upon further investigation into the issue, it was discovered that the electronic records were migrated without the original baselines set in the Macintosh environment. However, if the migrated data are auto-processed (baselines were automatically placed using pre-set criteria) using manually input data from the original MassChrom methods, then similar analyte results are obtained.

The major issue is therefore when quantifying data we are unable to comply with the full requirements of 21 CFR 11. However, there was no need to re-develop any method as similar results were obtained and being consistent with the comments of McDowall.[72]

Calibration curve parameters for original and converted data are shown in Table 31; the values are equivalent. However, the criteria chosen for acceptance of the data migration were based on the calculated results. As the analysis is based upon a comparative method of analysis (chromatography), the results were deemed the best way of evaluating if the conversion was successful. The key question is would the same decision be taken on the data? Therefore, a regression line of the MassChrom *versus* the Analyst across all concentrations should have a correlation co-efficient close to 1.0 if the results were the same by both methods. These data are shown in Figure 44.

23.10.4 History Logs

MassChrom does not have an audit trail associated with the data but it does have a history log associated with each data file that notes data and time of creation and

Table 30 *Comparison peak areas from MassChrom with the same data converted and calculated by Analyst*

Analyte standard concentration (ng mL^{-1})	MassChrom peak area	Analyst peak area
10	4366	4544
20	7851	8383
50	22,867	23,160
100	45,204	47,667
500	205,054	205,822
1000	399,296	401,330

Table 31 *Calibration curve parameters calculated by MassChrom and Analyst*

Calibration parameter	MassChrom	Analyst
Slope	0.00365	0.00362
Intercept	0.00127	−0.00036
Regression co-efficient	0.99726	0.9960

changes made to the data. The entries created in the Macintosh environment were migrated to the Analyst environment exactly and were updated following change of a baseline or similar events.

23.11 Data Migration Summary

When considering a data migration and system retirement project the following approaches are suggested. Think first and understand the complexity of the whole system and technical problems associated with it. This is important and whilst it will slow the overall project initially, will enable the actual work to proceed more smoothly than would be the case if this step were omitted.

You will be unlikely to solve the problem at the first attempt, therefore adopt an evolutionary approach to the issues. This is illustrated in this chapter where the number of scripts rose from 1 to a final 12. Do not rush into actions, therefore draw up a data migration plan and then do nothing for at least a week to enable you to review the plan critically and refine the approach: is it feasible and what is the regulatory risk?

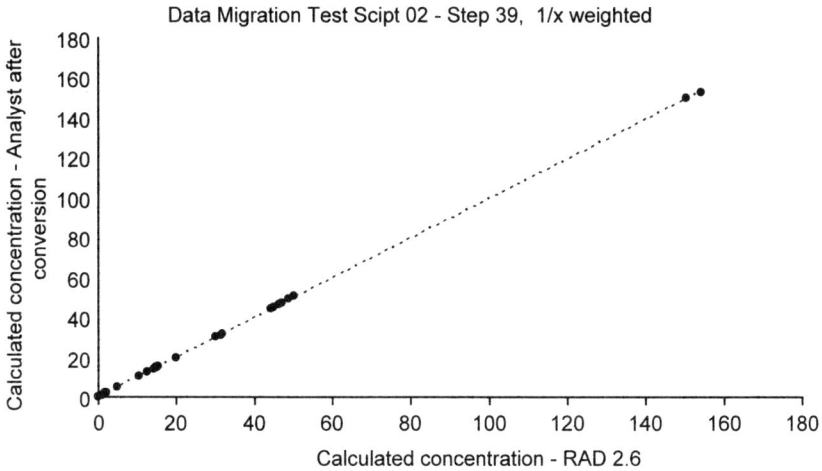

Figure 44 *Regression analysis of Macintosh and Analyst data showing equivalent results obtained from the data migration*

Be practical and flexible, as you will find unexpected issues when least expecting them. The better prepared you are the less likely these issues will be major and affect the data migration adversely.

Large volumes of data will be produced when validating the data migration process, plan well in advance how to capture and handle these data. These data will be both paper and electronic files, manage both well and have file-naming conventions.

CHAPTER 24

CDS System Retirement

The final stage in the life cycle of a CDS is the retirement of the components of the system.

24.1 What do the Regulators Want?

There is again nothing formal or implied in the GLP or GMP regulations about what is needed when retiring or decommissioning a chromatography data system or indeed any computerised system used in a controlled environment.

Often system retirement is a euphemism for either throwing the components out of the organisation or donating the items to an educational institution. There needs to be a more structured process where the components have documented evidence of their removal from operational service.

24.2 Generic Process for System Retirement

The overall process flow for system retirement is shown in Figure 45, the involvement of both management and quality assurance in the process is a critical success factor.

24.2.1 Notification of System Retirement

There needs to be a formal notification that the old CDS will be retired. This is important as there needs to be continuity and co-ordination between the old system being retired and the new CDS being installed, qualified and becoming operational. There can be no gaps in service support as chromatography in most laboratories is a major analytical technique. Therefore, there needs to be an orderly transition between the two systems and work that is ongoing is smoothly transferred to the new CDS.

Continuity of laboratory operations is important as there is typically never a good time to transfer between systems. This is particularly important to look at the impact of the new system on continuing operations; stability studies and other long-term work will start analysis on one CDS and finish on the new one. Work transfer needs to be planned and an assessment of any differences between the two systems understood and factored into the transition to the new system.

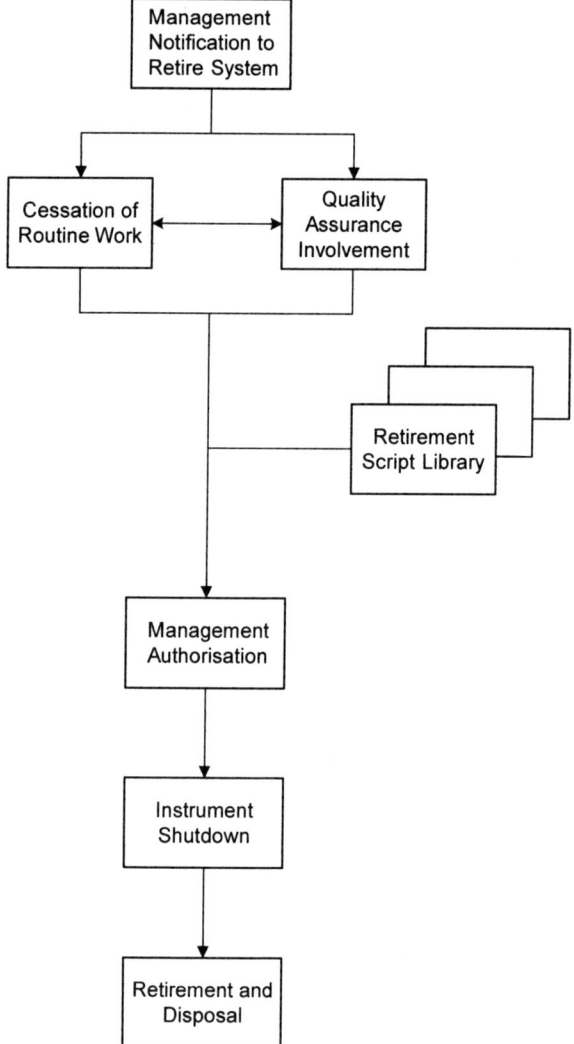

Figure 45 *Process flow for system retirement*

Management have a key role to play in this process, which will usually be complicated by the situation that a system may be operational across a number of departments or even sites.

24.2.2 Involvement of Quality Assurance and IT

There will be regulatory implications in the retirement of the old CDS and therefore quality assurance need to be involved to understand and discuss the implications of the retirement and underwrite them from a regulatory perspective. QA will also be involved with reviewing and approving the documents involved; therefore, they need to be on board early in the process so that the work does not come as a surprise.

In addition, other staff may be involved such as IT Department personnel; this is particularly important as there may be procedures within IT for the retirement and disposal of computer hardware from site, particularly, the means of ensuring that proprietary information is removed from the hard drives of servers and work-stations that may be retired and moved off site (*e.g.* reformatting the drives or physical removal and destruction).

24.2.3 Cessation of Work

At a planned date, work using the old system must stop to allow the rest of the retirement process to occur. Typical factors involved in planning this date are:

- The new system has been installed, qualified and released for operational use.
- Instruments are connected and qualified to use the new system.
- Methods and reports in the new CDS are operational and work.
- Users are trained to work with the new CDS system.
- Procedures for using the new system are effective.

To meet all these criteria the date for ceasing to work with the old system may be immediately after the last users have been moved to the new system. In other instances the cessation date may be 6 months after the last transition to accommo-date any need to access and interpret data acquired on the old CDS. There are a number of factors in this decision such as age of the computer hardware the old system uses, licences and maintenance agreements with the old CDS vendor and criticality of the work carried out by the system.

24.2.4 Shutdown of the System

When work has ceased on the system, the main retirement work can actually begin. This involves some or all of the following elements:

- Final backup of the CDS data followed by any migration to an on-line archive or SDMS as outlined in the retirement plan.
- Turning off the power to the system (network server and any data servers and A/D units); you can have an option at this time to disconnect these items from the network and reuse any IP addresses if required.

24.2.5 Documenting Retirement and Disposal

For the actual retirement and disposal you will need a series of approved test scripts to document the process, as you progress with different system retirements, you can develop a library of test scripts that can be taken and modified to undertake future system retirement tasks depending on the size and complexity of the system. The tasks that will make up the system retirement are:

- Collation and archive of the current application software disks and vendor paper documentation on how to use the system, *e.g.* user manuals, quick reference guide and system administrator's guide.
- Collation and archive of the configuration management, change control and operational log books for the system plus the original copies of the SOPs for these processes.
- If SOPs for the system are distributed as hardcopies, the authorised copies need to be retired from the laboratory and any other areas where the system is used. The original copy will be placed in the archive and the other copies destroyed. Alternatively, if you have electronic copies of SOPs, then each one will be made historical and withdrawn from active circulation in the EDMS.
- If a thick client application model is used, the software should be deleted from the workstations where it is installed. This can be a major part of the work of retirement. If the workstations using the system have been upgraded and replaced as part of the rollout of the new CDS, this process may have already been performed. Serving the application using a terminal emulation application such as Citrix® will entail much less work as only the viewer applet will be installed on workstations and this may be part of the normal desktop in an organisation and need not be part of the system retirement.
- Analogue to digital units and data servers of the old system will be removed from the laboratory and either discarded, donated to an educational institution or could even be part exchanged for the new CDS. There needs to be documented evidence of the disposal of these items.
- The CDS data server and disks will be removed from its rack in the data centre/computer room and the disks physically destroyed or reformatted and all components recorded as discarded.
- The current cycle of backup tapes need to be taken out of circulation and recycled or destroyed as appropriate.

The completed test scripts and documented evidence of actions should be reviewed by QA and then also archived.

24.3 Case Study of System Retirement

Under the data migration and system retirement test plan of the case study outlined in Chapter 23, three test scripts were written for the formal retirement of the obsolete mass spectrometry systems. As these systems were essentially of the same configuration, the test scripts were identical and just varied with the name and identification of an individual system.

The essence of each retirement test script was a *pro-forma* checklist for the systematic collection and confirmation of activities involved in retirement of an instrument. Sections within each test script for the retirement of a system included:

Table 32 *Some items for safety from an LC–MS system retirement checklist*

Safety checklist: Ensure the following tasks have been completed	Complete (yes/no)
External surfaces of the MS instrument have been cleaned with detergent and decontaminated	
Radioactive counts on the external surfaces of the instrument are < 10 counts above background	
Vacuum pump hoses have been disconnected from the instrument, coiled up and taped to the diffusion pumps	
All electrical cables have been disconnected from the wall sockets, the cables coiled up and taped to the instrument surface	

- *Component inventory.* All components of the system including the computer, network connections, software and MS instruments are listed in the test script (this is supplied from the system inventory and information gathering stages of the process outlined in Figure).
- *Data.* It was confirmed that all data have been backed up and then copied across to a server and have not been corrupted. This is followed by deletion of the data on the hard drive.
- *Computer.* Disconnection of the computer from the network and informing the IT department that the socket (IP address) can be reallocated if required. The hard drive of the Macintosh was reformatted before the computer was removed from site to ensure that no confidential data remained.
- *Mass spectrometer.* There were several stages to this where it was confirmed that the instrument was biologically and radiologically decontaminated before allowing it to be removed from the site. A portion of this checklist is shown in Table 32.
- *Finance.* The fixed asset numbers and identities of the components retired were passed to the Finance Department to update the asset register and show the item as decommissioned.

The format of a retirement test script is less structured than a PQ test script shown in Tables 21 and 22 with no expected results but documented evidence is collated and there are simpler acceptance criteria.

CHAPTER 25

Retrospective Validation

Systems that are operational in regulated environments but are not validated must be brought under control. This is termed as retrospective validation. Retrospective validation costs must be balanced against the cost of a replacement system as this is the most expensive way to validate a system.

25.1 What do the Regulators Want?

25.1.1 PIC/S Guidance

The best regulatory advice for retrospective validation comes from the PIC/S guidance document[31] where Chapter 16 is devoted to the subject; selected quotations are presented here, however the reader is strongly advised to read the whole section if faced with a retrospective validation of a CDS.

Section 16.1: Retrospective validation is not equivalent to prospective validation and is not an option for new systems. Firms will be required to justify the continued use of existing computerised systems that have been inadequately documented for validation purposes. Some of this may be based on historical evidence but much will be concerned with re-defining, documenting, re-qualifying, prospectively validating applications and introducing GXP related life-cycle controls ...

Section 16.2: A significant number of legacy systems may operate satisfactorily and reliably, however, this does not preclude them from a requirement for validation. The approach to be taken is to provide data and information to support the retrospective documentation of the system to provide validation and re-qualification evidence ...

Section 16.4: Nevertheless, the validation strategy would be consistent with the principles established for classic retrospective validation where the assurances are established, based on compilation and formal review of the history of use, maintenance, error report and change control system records and risk assessment of the system and its functions. These activities should be based on documented URS's. If historical data do not encompass the current range of operating parameters, or if there have been significant changes between past and current practices, then retrospective data would not of itself support validation of the current system.

Section 16.5: The validation exercise for on-going evaluation of legacy systems should entail inclusion of the systems under all the documentation, records and procedural requirements associated with a current system. For example, change control, audit trail(s), (where appropriate), data & system security, additional development or modification of software under a QMS, maintenance of data integrity, system back up requirements, operator (user) training and on-going evaluation of the system operations.

Section 16.6. Ultimately, regulated users have to be able to demonstrate:

- *defined requirements*
- *system description, or equivalent*
- *verification evidence that the system has been qualified and accepted and that GXP requirements are met*

25.1.2 Regulatory Requirements Summary

The key requirement is that unvalidated CDS systems are brought into a state of control, part of this is by reviewing historical performance but you will have to write or at least update the URS and test the system against those requirements.

25.2 Literature References to Retrospective CDS Validation

Guidance for retrospective validation in the literature tends to be presented in overview only. In the paper by Johansson *et al*,[74] a retrospective validation of a CDS is described in detail. The key difference between a prospective and a retrospective CDS validation is the gap and plan phase.

25.3 Gap and Plan for Retrospective Validation

The gap and plan phase is an essential stage in the retrospective validation of any computerised system. The process is shown in more detail in Figure 46.

The material in this Chapter is based on the work from the Johansson et al case study.[74]

25.3.1 Collect Existing Documentation

First stage in the retrospective validation of a CDS is all of the existing documentation on the system must be collected; this could include:

- validation plan
- URS
- documentation from the selection process
- purchase order and delivery notes
- qualification tests and documentation
- PQ plan and test scripts
- training materials
- operating manuals from the vendor
- standard operating procedures

In the case study,[74] the system was relatively new and most of the available documentation was retrieved, as documentation was easily available. Furthermore, the personnel operating the system have been involved with the project from the

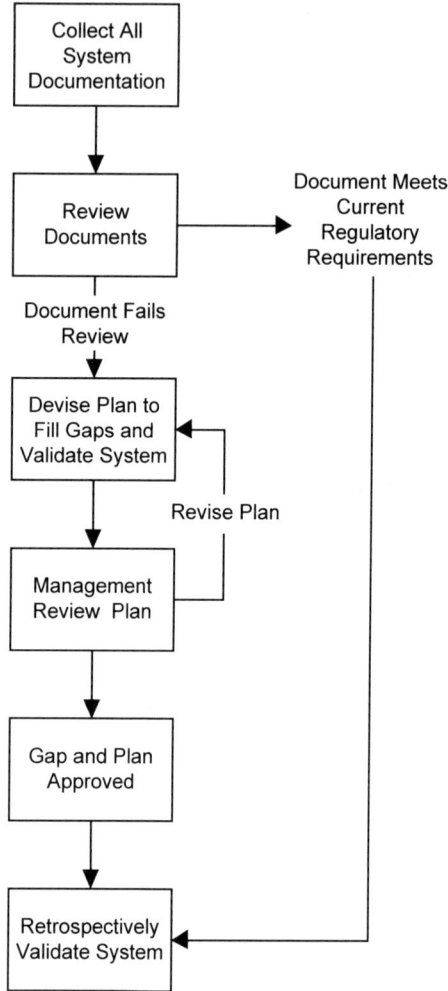

Figure 46 *Gap and plan for retrospective validation of a CDS*

start. This is in contrast to a system where documentation may be non-existent and personnel may have left the company or indeed the company has reorganised or merged.

When all the documentation has been collected, a list is made. This can be compared against the current regulatory regulations, industry guidelines and the corporate validation policy. This generates a list of missing documents and identified the gap to be filled.

25.3.2 Review Existing Documents

Next, the existing documentation must be reviewed to see that each item is of suitable quality, coverage and fitness for purpose. The mere existence of a document does not mean that its quality and coverage is good. Poor documents must be completed, or otherwise discarded and a new one written that meets the current compliance requirements. For instance, is there a current user requirements specification (URS)? Is it specific enough to allow qualification tests to be constructed? If a URS consists of one or two pages of general statements for a data system, such as:

- the data system performance must be fast
- user-friendly operation

This means that there is no firm requirement to allow a meaningful test to be constructed. The assessment of documents may result in more documents being added to the gap list.

25.3.3 Planning to Bridge the Gap

Once the gap has been defined, there must be a decision made to either write the key documents and fill the gap or for management to take the business risk not to write them. Time and resources must be included in this plan. This list of documents to be written, authorised by management, is the output of the gap and plan phase.

The gap and plan identified that there were several key documents that were required in the case study[74]; these were:

- validation plan
- user requirements specification
- test plan for the qualification of the system
- user test scripts (performance qualification)
- change control and configuration management SOP
- system description

The process for the retrospective validation is to write these documents and execute any PQ testing as necessary.

25.3.4 Management Underwrite the Plan

Estimates of the time and resources required to complete the documents and testing are drawn up and submitted to management for review and approval; in essence, there are only four options available:

- accept the plan as presented
- modify the plan to acceptable scope and cost
- reject the plan and continue to operate the unvalidated system
- reject the plan, retire the system and purchase a new system that will be more cost efficient to validate

The decision is usually based on a number of factors such as age of the system, available budget, cost of doing the work, the cost of getting caught if nothing is done and when a regulatory inspection is due. There is usually no simple or straightforward formula as each case tends to have unique factors to consider. However, operating an unvalidated CDS in a regulatory environment can be high risk as most inspectors know about chromatography data systems and ask to see evidence of the validation of the system.

Once the plan is agreed, you can use the appropriate chapters of this book to undertake the retrospective validation of your system.

CHAPTER 26

References

1 GAMP Forum, *Good Automated Manufacturing Practice (GAMP) Guide Version 4*, International Society for Pharmaceutical Engineering, Tampa, FL, 2001.

2 Institute of Electronic and Electrical Engineers, *Software Engineering Standards 2003 Collection*, Institute of Electronic and Electrical Engineers, Piscataway, NJ, 2003.

3 N. Dyson, *Chromatographic Integration Methods*, 2nd edn, Royal Society of Chemistry, Cambridge, 1998.

4 A. Felinger, *Data Analysis and Signal Processing in Chromatography*, Elsevier, Amsterdam, 1998.

5 R.D. McDowall, *LC–GC Eur.*, 1999, **12**, 568–578.

6 R.D. McDowall, in: *Analytical Chemistry in a GMP Environment – A Practical Guide*, J.M. Miller and J.B. Crowther (eds) Wiley, New York, 2000, 395–421.

7 C. Burgess, D.G. Jones and R.D. McDowall, *LC–GC Int.*, 1997, **10**, 791–795.

8 United States Pharmacopeial Convention Inc., *United States Pharmacopoeia 28*, United States Pharmacopeial Convention Inc., Rockville, MD, 2004.

9 European Pharmacopoeia, *European Pharmacopoeia*, Strasbourg, 2003.

10 C.L. Burgess, in: *Development and Validation of Analytical Methods*, C.L. Riley and T.W. Rosanske (eds), Pergamon Press, Oxford, 1996, 352.

11 R.J. Davis, in: *Development and Validation of Analytical Methods*, C.L. Riley and T.W. Rosanske (eds), Pergamon Press, Oxford, 1996, 352.

12 US Food and Drug Administration, *Fed. Reg.*, 1978, **41**, 45076.

13 US Food and Drug Administration, *Pharmaceutical cGMPs for the 21st Century: A Risk-Based Approach*, 2002.

14 US Food and Drug Administration, *Fed. Reg.*, 1978, **43**, 45077 and subsequent ammendments as listed on www.fda.gov.

15 US Food and Drug Administration, *Fed. Reg.*, 1996, **61**, 52601–52662.

16 International Conference on Harmonisation (ICH), *ICH Q7A – Good Manufacturing Practice for Active Pharmaceutical Ingredients (CPMP/ICH/ 4106/00)*, 2000.

17 US Food and Drug Administration, *Fed. Reg.*, 1997, **62**, 13430–13466.

18 US Food and Drug Administration, *Guidance for Industry: 21 CFR Part 11; Electronic Records; Electronic Signatures Part 11 Scope and Application*, 2003.

19 R.D. McDowall, *Am. Pharm. Rev.*, 2001, **4(2)**, 91–96.

20 GAMP Forum, *Good Practice Guide – IT Infrastructure Control and Compliance*, International Society for Pharmaceutical Engineering, Tampa, FL, 2005.

21 R.D. McDowall, *Am. Pharm. Outsourcing*, 2004, **5(3)**, 29–35.

22 US Food and Drug Administration, 2001.

23 US Food and Drug Administration, *Withdrawn Guidance for Industry: 21 CFR Part 11; Electronic Records; Electronic Signatures Validation*, 2001.

24 US Food and Drug Administration, *Withdrawn Guidance for Industry: 21 CFR Part 11; Electronic Records; Electronic Signatures Time Stamps*, 2002.

25 US Food and Drug Administration, *Withdrawn Guidance for Industry: 21 CFR Part 11; Electronic Records; Electronic Signatures; Maintenance of Electronic Records*, 2002.

26 US Food and Drug Administration, *Withdrawn Guidance for Industry: 21 CFR Part 11; Electronic Records; Electronic Signatures Electronic Copies of Electronic Records*, 2002.

27 Medicines Control Agency, *Rules and Guidance for Pharmaceutical Manufacturers and Distributors 2002*, 6th edn, The Stationary Office, London, 2002.

28 Organisation for Economic Co-operation and Development (OECD), *The Application of the Principles of GLP to Computerised Systems, Environment Monograph 116*, Organisation for Economic Co-operation and Development, Paris, 1995.

29 US Food and Drug Administration, *Guidance for Industry: General Principles of Software Validation*, 2002.

30 US Food and Drug Administration, *Guidance for Industry: Computerized Systems in Clinical Trials*, 1999.

31 Pharmaceutical Inspection Convention/Scheme (PIC/S), *Good Practices for Computerised Systems in "GXP" Environments*, PIC/S, Geneva, 2003.

32 US Food and Drug Administration, *Gaines Chemical Company 483 Observations*, 1999.

33 US Food and Drug Administration, *Glenwood Warning Letter (m2663n)*, 1999.

34 US Food and Drug Administration, *Genesia Scicor Warning Letter (m2819n)*, 1999.

35 US Food and Drug Administration, *Noramco 483 Observations*, 2001.

36 US Food and Drug Administration, *Cordis Warning Letter (g4601d)*, 2004.

37 Parenteral Drug Association, *J. PDA*, 1995, **49**, S1–S17.

38 US Food and Drug Administration, *Guidelines for Process Validation*, 1987.

39 B. Boehm, *Some Information Processing Implications of Air Force Missions: 1970–1980*, RAND Corporation, Santa Monica, CA, 1970.

40 D. Parriott, *LC–GC*, 1994, **12**, 132–135.

41 B. Boehm, in: *Euro IFIP 79: European Conference on Applied Information Technology*, P.A. Samet (ed), North-Holland, Amsterdam, 1979.

42 GAMP Forum, *Best Practice Guide – Laboratory Systems*, International Society for Pharmaceutical Engineering, Tampa, FL, 2004.

43 Society of Quality Assurance, *Computer Validation Initiative Committee (CVIC), Risk Assessment/Validation Priority Setting*.

44 International Standards Organisation, *ISO Standard 14971 – Medical Devices – Application of Risk Management to Medical Devices*, International Standards Organisation, Geneva, 2000.

45 US Department of Defense, *Military Standard (MIL-STD-1629a) Procedures for Performing a Failure Mode, Effects and Criticality Analysis*, US Department of Defense, Washington, DC, 1980.

46 Society of Automative Engineers, *Failure Mode Effect Analysis*, Society of Automotive Engineers, 1993.

47 R.D. McDowall, in: *Computer Systems Validation: Quality Assurance, Risk Management and Regulatory Compliance for Pharmaceutical and Healthcare Companies*, G. Wingate (ed) Interpharm/CRC, Boca Raton, FL, 2004, 465–510.

48 R.D. McDowall, *Am. Pharm. Rev.*, 2004, **7**, 20–25.

49 C. Kornbo and R.D. McDowall, *Sci. Comput. Instrum.*, 2002. January issue 12–14.

50 IEEE Standard 1012–1998, Validation and Verification Plans, Institute of Electronic and Electrical Engineers, Piscataway, NJ, 1998.

51 IEEE Standard 829–1998, Software Test Documentation, Institute of Electronic and Electrical Engineers, Piscataway NJ, 1998.

52 International Standards Organisation, *ISO Standard 9000: 2000 Quality Management Systems – Fundamentals and Vocabulary*, International Standards Organisation, Geneva, 2000.

53 International Standards Organisation, *ISO Standard 9001: 2000 Quality Management Systems – Requirements*, International Standards Organisation, Geneva, 2000.

54 International Standards Organisation, *ISO Standard 9002: Quality Assurance in Production, Installation and Servicing*, International Standards Organisation, Geneva, 2000.

55 International Standards Organisation, *ISO Standard 9003: Quality Assurance in Final Inspection and Test*, International Standards Organisation, Geneva, 2000.

56 International Standards Organisation, *ISO Standard 9003: 2004 Software Engineering – Guidelines for the Application of ISO 9001:2000 to Computer Software*, International Standards Organisation, Geneva, 2004.

57 Parenteral Drug Association, *Technical Report 32: Assessment of Supplier Quality Management Systems*, Parenteral Drug Association, 1999.

58 J. Moore, P. Solanki and R.D. McDowall, Chemometrics and Intelligent Laboratory Systems, *Lab. Automat. Inf. Manag.*, 1995, **31**, 43–46.

59 R.D. McDowall, in: *Computer Systems Validation: Quality Assurance, Risk Management and Regulatory Compliance for Pharmaceutical and Healthcase Companies*, G. Wingate (ed), Interpharm/CRC, Boca Raton, FL, 2004, 465–510.

60 W.B. Furman, T.P. Layloff and R.T. Tetzlaff, *JAOAC Int.*, 1994, **77**, 1314–1317.

61 M. Freeman, M. Leng, D. Morrison and R.P. Munden, *Pharm. Technol.*, 1995, **10**, 45–48.

62 C. Burgess, D.G. Jones and R.D. McDowal, *Analyst*, 1998, **123**, 1879–1886.

63 US Food and Drug Administration, *Spolana Warning Letter (m4314n)*, 2000.

64 M. Fewster and D. Graham, *Software Test Automation – Effective Use of Test Execution Tools*, Addison-Wesley, Harlow, 1999.

65 Institute of Electronic and Electrical Engineers, *Software Engineering Standards*, Institute of Electronic and Electrical Engineers, Piscataway NJ, 1998.

66 US Food and Drug Administration, *Compliance Policy Guide (CPG) 7132a.08: Computerized Drug Processing – Identification of "Persons" on Batch Production and Control Records*, Food and Drug Administration, Washington, DC, 1982.

67 S.H. Gambie, D.L.M. Weller and P. Withers, *Definition of Raw Data*, British Association of Research Quality Assurance (BARQA), 1994.

68 US Food and Drug Administration, *Inspection of Pharmaceutical Quality Control Laboratories*, Food and Drug Administration, Washington, DC, 1993.

69 International Conference on Harmonisation (ICH), *Electronic Common Technical Document (ICH M2 EWG)*, International Conference on Harmonisation, Geneva, 2003.

70 J. Liscouski, *Laboratory and Scientific Computing – A Strategic Approach*, Wiley, New York, 1995.

71 D. Browne, T. Thompson, D. Mole and R.D. McDowall, *J. Validat. Technol.*, 2002, **8**, 250–259.

72 R.D. McDowall, *LC–GC Eur.*, 2000, **13**, 35–38.

73 D. Browne, T. Thompson, D. Mole and R.D. McDowall, *LC–GC Eur.*, 2001, **13**, 687–692.

74 P. Johansson, B. Wikenstedt and R.D. McDowall, *LC–GC Eur.*, 1999, **11**, 88–96.

Subject Index